PIECEWISE LINEAR MODELING AND

PIECEWISE LINEAR MODELING AND ANALYSIS

by

DOMINE M. W. LEENAERTS
and
WIM M. G. VAN BOKHOVEN

Eindhoven University of Technology

KLUWER ACADEMIC PUBLISHERS
BOSTON / DORDRECHT / LONDON

Library of Congress Cataloging-in-Publication Data

ISBN 0-7923-8190-4

Published by Kluwer Academic Publishers,
P.O. Box 17, 3300 AA Dordrecht, The Netherlands.

Sold and distributed in North, Central and South America
by Kluwer Academic Publishers,
101 Philip Drive, Norwell, MA 02061, U.S.A.

In all other countries, sold and distributed
by Kluwer Academic Publishers,
P.O. Box 322, 3300 AH Dordrecht, The Netherlands.

Printed on acid-free paper

Printed in the Netherlands

To Lisanne for here love and support

D.M.W. Leenaerts

CONTENTS

PREFACE

Confronted with solving a nonlinear (static) equation people often use a very old technique. One simply approximates the nonlinear behavior with one or several linear equations. By nature we have more feeling for linear equations, we do know more about linear algebra than nonlinear system theory and we often can solve the problems by hand. If the linear approximation is accurate enough, the solution will be close to the exact solution. Otherwise one can either approximate more carefully the nonlinear behavior, resulting probably in a more complex linear problem or one can use the approximated solution as a starting point for solving the nonlinear equation. If one approximates a certain nonlinear behavior by a set of linear equations, the set is called a *piecewise linear approximation* of the nonlinear behavior. In case we have a nonlinear function, the linear approximation is called a piecewise linear function.

Certainly when more linear segments are necessary in approximating nonlinear behavior, one is searching for a method to be able to handle all these segments easily. Furthermore, it would be of interest to group them such that the original problem could be solved with certain advantages with respect to e.g. complexity, data storage,... even when a computer would still be necessary to solve the problem. For these reasons people were looking for a certain *model description.* This model description should be compact, i.e. a minimum of information should be stored to reproduce the piecewise linear behavior. The description should also allow for simple computation, the linear segments should be easily be extracted from the model. Or in other words, given the input for the model, the output should be obtained without to many difficulties. Finally, the model should make it possible to store as many different piecewise linear functions as possible. With different we refer to the dimension of the function, but also to the type of the function, e.g. one-to-one, one-to-many or many-to-one,...

Simulation programs play an important role in the design of integrated electronic systems. As an IC-prototype development is still very expensive, it will be of great advantage to have means available to prevent this money to be spilled because of some design error or unexpected circuit behavior. The on chip density of transistors nowadays allows to design mixed signal circuits on a single wafer. This means that we can combine analog and digital circuitry in one large system. To handle this complexity, the simulation program should be able to perform the analysis on a higher level than the basic transistor level and should be able to analyze a mix of parts of the system on transistor level with other parts on behavioral level. This means that the program must handle all kinds of models and hence all kinds of solution techniques. It would be to advantage if any different behavior (e.g. analog, digital, behavioral) would result in models which all have the same data format. Because only then one solution algorithm would do. Piecewise linear models do have this property. Their data format is exactly the same for any kind of function and hence piecewise linear techniques can be of help to analyze a circuit behavior certainly when the circuit possesses more than one operating point or a transfer characteristic which consists of several disjunct parts.

This book tries to guide the reader through the world of piecewise linear techniques ranging from modeling aspects to analyzing piecewise linear circuits. During writing an attempt is made to cover all issues without trying to be complete and cover all subjects in detail. Helpful examples are thoroughly discussed and a large list of references can help the reader on its way if more detailed background information is needed.

The book follows a logical trace and starts in chapter 1 with the link between electrical networks and piecewise linear modeling of such networks. It is shown how a network containing ideal diodes can be transformed into several different model descriptions. This results finally in the state model description as proposed by van Bokhoven.

Chapter 2 discusses in detail all well known model descriptions so far, the models presented by Chua, Kahlert, Güzelis, van Bokhoven and Leenaerts. Explicit and implicit model descriptions will be treated and it will be shown that the relation between these two is based on the modulus transformation. By then it must be clear which model can be used and which not for a certain given piecewise linear function and what the limitations of a certain model description are. It will be shown that implicit model descriptions have advantages compared to explicit descriptions related to the class of functions they can handle.

For a given input vector one needs a solution algorithm to obtain the output vector in case of an implicit model description. The underlying problem is known as the Linear Complementary Problem (LCP) and is the subject of chapter 3. Several solution algorithms to solve the LCP do exist and they will be discussed on the hand of one and the same example to show their advantages and drawbacks. The most familiar methods are those proposed by Katzenelson and Lemke. However also a powerful method, based on the algorithms of Tschernikow, will be discussed. This latter method, as will be shown later, can be used to find multiple solutions of the LCP.

Having the model description to describe the network elements and a solution algorithm to solve the LCP, in chapter 4 it will be shown how these two items can be combined to develop a piecewise linear analysis tool, able to solve piecewise linear circuits and systems. Aspects as DC, AC and transient simulation will be treated but will also deal with the DC bias point and how to perform a hierarchical analysis. The discussed algorithms are developed for the PL simulator PLANET, but other known PL analysis tools will be discussed as well

Chapter 5 explains the issue how to find a piecewise linear function for a certain behavior. For example, how to model a digital component like a NAND function or what does a piecewise linear model of a MOS transistor looks like. With references to basic electrical elements, but also to rather complex circuits the reader will be explained how to find the PL models. Some simulation results of real applications will be given to demonstrate the power of piecewise linear modeling compared to the approach as in SPICE-like simulators. Two programs will be discussed which automate this modeling process up to a certain level.

Finally, chapter 6 focuses on the problem of multiple operating points and driving point characteristics having several unconnected curves. Many circuits and systems posses several distinct operating points and the question is how to find them all and fast. It turns out that the problem is inherently complex and hence there is no simple way to solve it. To find all solutions one has to go through the complete domain space and the differences in the methods lie in the way how to scan this space, using knowledge of the model description.

The book is written after gaining experience in this field for more than 10 years. Discussions and conversations with colleagues are too numerous to list in detail but we must acknowledge particular debts to a few persons. We would like to thank dr. Tom Kevenaar from Philips, who participated for a long time in this research. He developed and coded main parts of the PL simulator PLANET and contributed in the extension of the explicit model descriptions. Many thanks also to dr. Leon Chua from U.C Berkeley for the many fruitful discussions in the field of model descriptions and in trying to find the ultimate model that can cover any piecewise linear model. We had many stimulating discussions with dr. Joos Vandewalle and his (former) Ph.D. students from KU Leuven on the Linear Complementary Problem and on multiple solutions of (PL) equations. Furthermore we would like to thank our former Ph.D. students dr. W. Kruiskamp for developing several PL models of electrical elements and dr. P. Veselinovic for developing the automatic model generator. Also thanks to Mrs. L. Balvers for drawing most of the pictures in this book.

March 1998

Domine Leenaerts
Wim van Bokhoven

CHAPTER 1

THE STATE MODEL OF A PIECEWISE LINEAR NETWORK

In this chapter we will discuss the development of a state model for a piecewise linear function from a network point of view. To this purpose we will introduce the ideal diode, as simplest piecewise linear model which can serve as the base of any other piecewise linear element.

1.1.   Model descriptions

The creation of a mathematical description that approximates the system's functionality is called *modeling* and the description itself the *model description* or simply the *model*. Approximating a nonlinear function or behavior by a known nonlinear mathematical function is an often used technique. The purpose is to get more insight in the actual behavior by applying known mathematical theorems on the mathematical function. In principal there are three main concepts in approximation:

- *Polynomial functions.* In the past many research has been done in the field of approximating functions using polynomials as well as transcendental functions. Applying these techniques finally led to simulators like SPICE. Widely known are the Taylor, Chebyshev and Padé approximations that use (rational) polynomial functions for the fitting [1]. In transcendental approximation, sinusoidal forms are used to approximate the nonlinear behavior.
- *Table-look-up.* By sampling the nonlinear function, a set of points is obtained which can be stored in a table. If a function value is needed, the corresponding entry in the table is used. However, if the function value is not directly stored, interpolation and extrapolation techniques are used to obtain the value. The advantage of such a method is its speed. This method is superior compared to function evaluation. In contrast however, the need of memory to store the

function can be extreme large. Furthermore it is not possible to gain insight in
the nonlinear behavior.
- *Piecewise linear functions.* Approximating the nonlinearity by using several
linear affine descriptions is a very old and often used method. By doing this,
the basic problem is transformed from a single nonlinear equation into several
linear equations. In mathematics, there is much more knowledge and flexibility
in linear theory than in nonlinear function theory. The accuracy of the solution
in this situation depends on the fineness of the approximation.

Certainly the relation with the linear algebra makes the PL modeling technique
of interest in the nonlinear network theory. This aspect will here be treated as we
will show that using ideal diodes the nonlinear network can be transformed into
several sets of linear sub networks. The diodes will tell us which linear sub network
must be considered for a given input. Now by applying modified nodal analysis
(MNA), each linear sub network can be described by $r = Ae$, with e the node
voltages and r the independent sources. For n diodes, representing 2^n sub
networks, there are 2^n MNA equations to store. They together describe a piecewise
linear function behavior. The description is called a *piecewise linear model
description.* However, straightforward storage of the mappings will demand a huge
storage capacity and therefore one searches for a model description such that all
those linear mappings could be stored with the minimum amount of data.
    Here we will show how, from a network point of view, one can develop a model
description that can store all the network information in a closed compact way with
a minimum of parameters.

1.2.   The ideal diode

When confronted with the question to develop a piecewise linear model for
nonlinear components in electrical circuits one obviously starts to look for the most
simple extension to the well-known linear components like resistors and linear
controlled sources. This extension should in one way supply a kind of *basic*
nonlinearity but in another way this nonlinearity should be as simple as possible,
expecting to extend this approach to more general nonlinearities later on.
    The first element that will come up satisfying those conditions seems to be the
semiconductor diode. It surely is one of the most simple nonlinear elements and it
has been used in the past for a long time already to synthesize or reproduce
nonlinear transfer functions in analog computers by realizing piecewise continuous
approximations. In Fig. 1.1 an example is shown of such piecewise linear
approximation. Of course the exponential current-voltage characteristic of the actual
diode will not yield a true PL function as can be seen from the output of the circuit,
but if designed carefully, the departure from the ideal behavior is only observable in
a slight rounding at the interval boundaries.

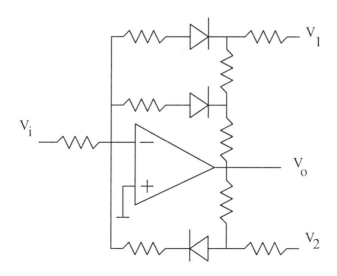

Fig.1.1. An analog piecewise linear approximation circuit

Taken this into account one can try to idealize the behavior of the diode. The *ideal diode* draws no reverse current when polarized into reverse bias and does not need any forward voltage to conduct an arbitrary large forward current. Such an idealization yields a V-I relation that exists of only two branches, one with *V=0* and *I>0* and one with *V<0* and *I=0* as depicted in Fig. 1.2. For reasons of symmetry we will reverse the voltage reference polarity of the ideal diode with respect to the normal convention such that the characteristic now reads

$$V > 0, I = 0 \text{ and } V = 0, I \geq 0$$

or equivalently

$$V, I \geq 0 \text{ and } V \cdot I = 0 \tag{1.1}$$

(see also Fig. 1.2).

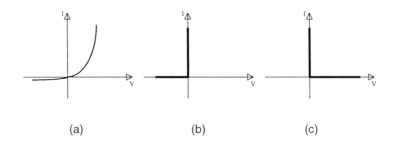

Figure 1.2. Diode characteristics with a) the actual behavior, b) the ideal electrical and c) the ideal behavior as defined in (1.1)

Note that the characteristic of the ideal diode can also be considered as piecewise linear by itself with 2 being the minimum number of PL-segments necessary to differentiate from fully linear elements. Thus indeed this diode seems very *basic* in this respect.

Furthermore, in any actual electrical network application, this element can only exist in one of two possible states - it *either* conducts with zero voltage, representing a closed connection between its terminals *or* it blocks the current in the reverse mode, behaving as an open circuit. This means that any linear circuit containing ideal PL-diodes only changes its topology when the diodes switch from conducting state i.e. switch from short- to open-circuit or the other way around. Therefore the response will remain linear in any conducting state of the diodes, but for different conduction state the response will be different since we deal with a network with switches that can change the topology. As the switching occurs in the point $V=I=0$, the response will automatically be continuous in the applied excitations. This property is very essential and will be used to advantage in the context of network analysis. We can apply methods from linear algebra which are more powerful and familiar to us than those from the nonlinear counterpart.

From now one we will only use and talk about ideal diodes when we mention the term diode unless explicitly stated otherwise.

1.3. Electrical networks and PL models

Let us consider a fairly simple network in which a number of resistors, independent voltage sources (batteries) and diodes are connected in parallel as given in the circuit of Fig. 1.3.

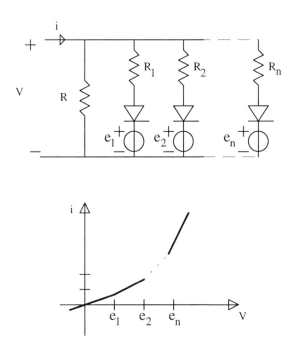

Figure1.3. Circuit example with ideal diodes and its *V-I* characteristic

In each parallel branch, the ideal diode starts to conduct when the input voltage exceeds the voltage of the battery. Suppose that the independent voltages are considered to increase in voltage, i.e. $e_1 < e_2 \cdots < e_n$. This means that an increasingly higher voltage is required at the input to include the parallel branches that are placed more to the right in the figure. However, when these branches start to conduct the total resistance is decreasing and hence the slope of the current voltage characteristic is increasing, leading to a monotone increasing *V-I* characteristic as also depicted in Fig. 1.3. Should we however allow resistors with negative values, then the slope of the characteristic can take any value and hence at least theoretically the proposed circuit can realize a one-dimensional piecewise linear relation between two variables (voltage and current at the input terminals). If we are able to describe the electrical behavior of the network of Fig. 1.3 we will have obtained a mathematical description of a one-dimensional PL function without any further restrictions. Should this network description result in an explicit solution, then this would yield an explicit PL-function, however it will always produce at least an implicit description. We will come back on this topic in chapter 2.

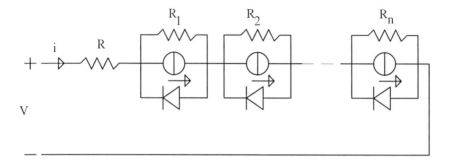

Figure 1.4. The dual circuit for the example in Fig. 1.3.

There we will derive in a mathematical way the explicit and implicit PL model descriptions and show their relation among each other.

From arguments from electrical network theory, we know that it is possible to construct the dual electrical network that produces the same functional relation with the roles of current and voltage interchanged (note that the ideal diode characteristic is self dual). That is, we can construct another topology given in Fig. 1.4 that yields comparable functional relations, but will result in a different description. Hence we immediately conclude that the description that we are looking for will not be unique.

In the above situation it is fairly easy to produce an explicit description of the v-i relation at the input terminals using basic mathematical functions. To this purpose consider the following expression

$$\lfloor x \rfloor = \frac{1}{2}\left(x + |x|\right) \tag{1.2}$$

that realizes a *ramp function* with the breakpoint at $x=0$. We will use this function in chapter 2 to develop a PL model description valid for a large class of PL functions. Based on our previous discussion, using (1.2) the current in branch k satisfies for $k>0$

$$i_k = G_k \lfloor v - e_k \rfloor \text{ with } G_k = 1 / R_k \tag{1.3}$$

Application of (1.3) and summation over all branches immediately yields

$$i = Gv + \sum_{k=1}^{n} G_k \lfloor v - e_k \rfloor$$

or

$$i = Gv + \frac{1}{2}\sum_{k=1}^{n} G_k (v - e_k) + \frac{1}{2}\sum_{k=1}^{n} G_k |v - e_k| \tag{1.4}$$

or in a more general mathematical expression

$$y = Ax + b + \sum c|Dx + e|$$

that, as we will see in chapter 2, is the basic model description as proposed by Chua and Kang and is known as the sectionwise PL-representation [3].

1.4.   The states in a PL-mapping

The above derived model description is not as general as we wants. This is simply due to the chosen network, which is rather straightforward. However, we can consider the circuits in Fig. 1.2 and 1.3. as special cases of a linear memoryless electrical multiport network which is loaded at some of its ports by the previously defined ideal diodes.

   The network may contain resistors and fixed and controlled sources and, for later convenience, all of its ports are partitioned in two different sets, port-set 1 and port-set 2 (see Fig. 1.5).

   Let us assume that port-set 1 contains m ports and port-set 2 contains k ports. For port-set 2 all its ports are loaded by ideal diodes, thus the voltage u over the diodes and the current j through the diodes are k-dimensional vectors, i.e. $u, j \in R^k$. Then, based on the previous discussion, the conducting states of the diodes will depend on the m currents and m voltages at port-set 1.

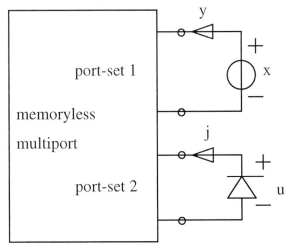

Figure 1.5. A memoryless electrical multiport loaded at some ports with ideal diodes

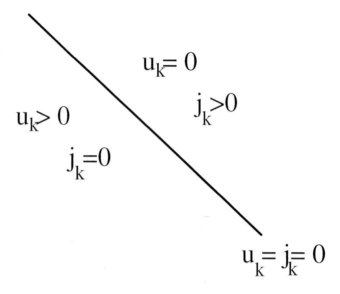

Figure 1.6. Separation of the space into two half spaces by the diode

We assume (without loss of generality) that the port-set 1 is excited by voltage sources. In the light of the PL mapping that we are required to construct, we map the port variables at port-set 1 on new vectors $x, y \in R^m$ according to $x=V$ and $y=I$.

Consider diode k that has only two states that are separated by the condition $u_k j_k = 0$. The diode separates also then the space spanned by x and y into two half-spaces, one in which the diode conducts and one for which it blocks (see Fig. 1.6).

The boundary between the two half spaces is a hyperplane determined by $u_k = j_k = 0$. Since in any conducting state the response of the network will remain linear in terms of the applied voltage excitation, the components of the vectors y, u and j are all linear relations in the components x_i of the vector x. Hence the hyperplane can be rewritten such that a linear combination of the components x_i of x are equal to zero, i.e.

$$c^t x + g = 0 \qquad\qquad (1.5)$$

Based on this discussion, we may state that all the diodes together separate the complete input space into 2^k regions, called polytopes. Within each polytope, all diodes remain in one of their conducting states, some will conduct other will block. Within each polytope, we have a linear relation between x and y. Crossing a hyperplane, means that the diode, corresponding to this hyperplane, changes its state and hence we have an other topology, defined by the polytope in which we enter after crossing the boundary. Again we are confronted with a linear network.

The questions that remain are how the topology will change after crossing a boundary and are we able to find a close format description.

1.5.   The partitioning of the state space

Consider the situation that we have k diodes or hyperplanes, and therefore 2^k polytopes. A polytope will be defined by K_m and for each polytope we have a linear mapping representing the topology of the network for that polytope,

$$y = A_m x + f_m \qquad m = 1,2,\ldots,2^k \tag{1.6}$$

Clearly the complete set of (1.6) describes a PL mapping which is defined on a collection of polytopes. The boundaries of each polytope K_m will be formed by a set of bounding hyperplanes H_m^i according to

$$H_m^i = \left\{ x \middle| C_m^i x + g_m^i = 0 \right\} \tag{1.7}$$

that is a generalization of (1.5). Equation (1.7) defines a collection of half spaces V_m^i given by

$$V_m^i = \left\{ x \middle| C_m^i x + g_m^i \ge 0 \right\} \text{ and } K_m = \bigcap_i H_m^i \tag{1.8}$$

(see Fig. 1.7).

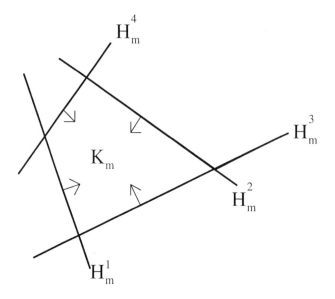

Figure1.7. Collection of hyperplanes H_m^i defining a polytope K_m as the

intersection of the half spaces V_m^i

The row vectors C_m^i in (1.8) are the normal vectors on the hyperplanes that bound the polytope K_m and point into the inward direction. They all can be considered as rows of a matrix C_m such that the polytope K_m is equivalently given by

$$K_m = \left\{ x \middle| C_m x + g_m \geq 0 \right\}$$

(1.9)

Thus the partitioning of the domain space of x is defined by 2^k relations of type (1.9) and in general would require the storage of 2^k matrices C_m and vectors g_m. The definition of the mapping on these polytopes would according to (1.6) in addition require the storage of 2^k sets of matrices A_m and vectors f_m.

In this way the total amount of storage to define a PL mapping and hence to describe a PL network could become exponential large, even for low dimensional spaces. Most often the mappings can however be modeled by fewer data, due to a special partitioning of the definition space.

To this purpose define each polytope K_m by specifying on which side of each hyperplane H^i it is situated. Note that for the hyperplanes the subscript m is removed to express that we only have a single set of hyperplanes H^i, $i \in \{1,\dots,k\}$ which can partition the space into a maximum of 2^k polytopes. This exactly fits with our network with ideal diodes. The k diodes can define at maximum 2^k topologies into which the network can be split up. Therefore, for any polytope K_m and each hyperplane H^i we either have one of two possibilities

$$C^i x + g^i \geq 0 \quad \text{or} \quad C^i x + g^i < 0 \tag{1.10}$$

Because the normal vectors on the hyperplanes were considered as rows of C_m and thus also for C^i we may collect all those normal vectors into a single matrix C. The same property holds for the vector g.

1.6. Properties of a PL mapping

Once the partitioning of the space is given, the various matrices A_m and vectors f_m in (1.6) also have to be defined in accordance with (1.10). We are finally looking for a compact description of the piecewise linear function as defined by the network with its ideal diodes. From a network point of view it is clear that the network is continuous and hence the underlying piecewise linear function. At that moment the matrices and vectors in (1.6) become related and may not freely be chosen.

To determine this dependency relation assume that the polytopes K_i and K_j are adjacent and have a common boundary hyperplane H_p defined by

$$H_p = \left\{ x \mid C_{p\bullet} x + g_p = 0 \right\} \tag{1.11}$$

with $C_{p\bullet}$ representing the p-th row from C, i.e. the normal vector corresponding to hyperplane H_p.

For each x on this hyperplane the mappings on both side of the boundary have to be identical, giving,

$$\forall_{x \in H_p} \ A_i x + f_i = A_j x + f_j \tag{1.12}$$

or

$$\left(A_i - A_j \right) x = b_j - b_i \tag{1.13}$$

Since A is of dimension R^{nxn} and x on the boundary belongs to an $(n-1)$-dimensional subspace this equation requires that $\Delta A = A_i - A_j$ is of rank one. By substitution of (1.11) in (1.13) one can easily find the relation

$$A_i = A_j + \frac{\left(b_i - b_j\right)C_{p\bullet}}{g_p} \tag{1.14}$$

that fully determines the relation between two mappings from adjacent regions. Relation (1.14) in certain sense describes the consistent variation property that will be discussed in more detail in section 2.1. Therefore the piecewise linear function is described completely by relation (1.14) together with the description of the state space.

Note that the above description yields a data complexity that grows polynomial with the number of hyperplanes, where as the number of possible topologies grows exponentially with the number of ideal diodes in the network.

It can be possible that two or some polytopes may overlap because they are situated on the same side of their common boundary hyperplane. For these cases it can be shown that the determinants of the corresponding matrices A have opposite sign. In addition the mapping which is represented by the description is then many to one and depending on the given affine mappings for each polytope, might be many to many as well. In any case, only one single matrix A_i has to be stored together with all vectors b and f and the matrix C. How this can be done in an efficient way will be explained in the subsequent section.

It is obvious that some algebraic mechanism will be needed to be able to use the PL mapping in an efficient way. The storage and updating of the description of the mappings as well as the calculation of the mapping itself can then be performed by standard operations from linear algebra. In chapter 3 several mechanisms to easily obtain the correct mapping for a given input x will be discussed while in chapter 4 the storage and updating of the description will be treated, when the design of a simulator to analyze piecewise linear networks is explained in detail.

1.7.   The complete PL model description

Consider again the network of Fig. 1.5 from which we learned that its response is a piecewise linear function that could be used to derive a closed form expression for a piecewise linear mapping. We recall from the V-I curves of the ideal diodes as given in (1.1) that for each diode at port-set 2 we have

$$\forall_{i\in\{1,\ldots,k\}} j_i \geq 0, u_i \geq 0, u_i \cdot j_i = 0$$

that can be written in vector notation

$$u, j \geq 0 \quad u^T j = 0 \tag{1.15}$$

with the inequalities taken component wise.

Furthermore we assume that the electrical behavior of the network within the solid box at its outside ports can be described by a port-admittance matrix H, resulting in

$$\begin{pmatrix} I_1 \\ I_2 \end{pmatrix} = \begin{pmatrix} H_{11} & H_{12} \\ H_{21} & H_{22} \end{pmatrix} \begin{pmatrix} V_1 \\ V_2 \end{pmatrix} + \begin{pmatrix} f \\ g \end{pmatrix}$$

Renaming I_1 and V_1 into y and x (see section 1.3) and the variables of port-set 2 u and j (because they are related to the diodes) and substituting (1.15) yields

$$y = Ax + Bu + f \tag{1.16}$$

$$j = Cx + Du + g \tag{1.17}$$

$$u, j \geq 0 \quad u^T j = 0 \tag{1.18}$$

which is known as the *state model* of a PL mapping $f : x \rightarrow y$ [2].

Equation (1.16) determines the *input-output mapping* of x onto y. The remaining two equations determine the *state of the mapping* from the electrical state of the ideal diodes. These diodes form a kind of *state variables* which, together with the input vector x determine the output y comparable to the situation in a state space model of a linear dynamic system. The conditions (1.18) are called the *complementary conditions* and u and j are complementary vectors. We will return to this issue and the *linear complementarity problem* in chapter 3.

Let us consider in some more detail the implications of this description in electrical network terms. As each diode can either be conducting or non-conducting the total number of different network topologies is at most 2^k with k the number of diodes. This means that there are at maximum 2^k different linear relations between x and y can be obtained. We indicate this by saying that the PL network can be in 2^k different states which each state determined by the conducting state of all k diodes.

In network terms, (1.17) is just one of the possible hybrid descriptions of the part of the circuit which is connected to the set of ideal diodes. By interchanging the port variables of port set 2 over all possible combinations we can come up with 2^k different hybrid descriptions equivalent to (1.16-1.18). These interchanging of variables can be performed by a Gauss-Jordan elimination step (so called *pivoting*) on the diagonal elements of the matrix D. In chapter 4 we will explain this pivoting process in more detail.

Of course the Gauss-Jordan elimination step will influence the other entries of A, B and f which in fact is precisely responsible for the changes in the linear mapping on the various polytopes as discussed in section 1.5. With k of such pivoting elements we have 2^k different pivotisations.

A basic representation of a mapping can be derived from (1.16) by forcing u equals zero, yielding $y=Ax+f$. An example of such a PL network is given in Fig. 1.8 for which the admittance matrix of the 4-port in the r.h.s. schematic reads

$$
\begin{pmatrix} I_1 \\ I_2 \\ I_3 \\ I_4 \end{pmatrix} = \begin{pmatrix} 6 & 2 & 1 & 2 \\ 2 & 2 & 0 & 0 \\ 1 & 0 & 1 & 0 \\ 2 & 0 & 0 & 2 \end{pmatrix} \begin{pmatrix} V_1 \\ V_2 \\ V_3 \\ V_4 \end{pmatrix} + \begin{pmatrix} -10 \\ -2 \\ -2 \\ -6 \end{pmatrix}
$$

This matrix then results in the following state model

$$
y = 6x + \begin{pmatrix} 2 & 1 & 2 \end{pmatrix} u - 10
$$

$$
j = \begin{pmatrix} 2 \\ 1 \\ 2 \end{pmatrix} x + \begin{pmatrix} 2 & 0 & 0 \\ 0 & 1 & 0 \\ 0 & 0 & 2 \end{pmatrix} u + \begin{pmatrix} -2 \\ -2 \\ -6 \end{pmatrix}
$$

$$
u, j \geq 0, u^T j = 0
$$

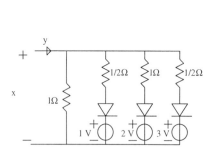
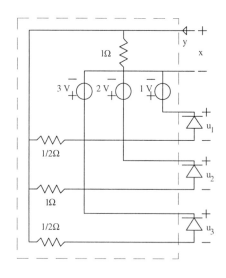

Figure 1.8. Example of a network with the given topology (left) that is transformed into a topology from which the state model can be obtained immediately (right).

It is possible to obtain the state model directly from Fig. 1.8b. Suppose that all diodes conduct, then the input-output relation yields

$$y = x + 2(x-1) + 1(x-2) + 2(x-3) = 6x - 10$$

which is the linear equation in the input-output mapping. The state equations are completely determined by the branches in which the diodes are located. For example port-set 2 conducts when x is larger than 1, hence

$$j_1 = x + u_1 - 1$$

and similar relations can be obtained for the other state equations.

We now only need the B-matrix to complete the state model. Each entry in this matrix can be found directly as the conductance in the branch with a ideal diode. For example, entry $b_3 = G_3 = 2$. In this way the complete state model can be obtained.

An important observation from the description is that apart from the complementary condition on the pair u, j the relation between all variables is *completely linear.* **This means that, keeping the complementarity as an obligatory side condition, the state model allows us to handle nonlinear piecewise linear systems as if they were linear systems.** The advantage of this property is obvious.

In chapter 2 we will see that the state model compared to other model descriptions covers the largest class of PL functions. In other words, so far we know it is the most general model description to model continuous piecewise linear functions.

1.8.   Conclusions

We showed that using a very simple piecewise linear element, in the form of an ideal diode, a model description can be developed. This piecewise linear model, the state model, contains in a compact manner all the possible topologies that a network can shows when it contains these ideal diodes. In a similar way, piecewise linear functions (the generalization of piecewise linear networks) can be modeled in this way as we will see in the next chapter.

We recall that this is only one of the possible model descriptions. There exist several possibilities that will be treated in the following chapter. We will also show their relations among each other.

CHAPTER **2**

PIECEWISE LINEAR MODEL DESCRIPTIONS

A piecewise linear function can be represented in many different forms. Since the end of the seventies with the presentation of the first model description for PL functions by Chua and Kang, many effort is put in a search for that particular explicit model description in which all continuous PL functions could be modeled. This under the restriction that there must exist a one-to-one relation between the model parameters and the PL function. Finally we want to obtain the model description that is able to handle all continuous PL functions. In this chapter we will give an overview of the state of art in this field. In contrast to the previous chapter where we developed the model from a network point of view we will here start from a mathematical background.

2.1. Fundamentals and definitions

All the PL functions to be considered are continuous functions, describing the mapping:

$$f: R^n \to R^m; x \to f(x)$$

A PL function consists of a collection of linear mappings, for each segment of the function exactly one. Each mapping is only valid in a certain subspace called a *polytope* which is a convex polyhedron in R^n. Such a polytope is bounded by a set of linear manifolds, called hyperplanes, with each hyperplane defined by

$$a^T x + b = 0$$

This situation is depicted in Fig. 2.1. Here we have four regions, $R_1, \ldots, R_4$ with the Jacobians $J_1, \ldots, J_4$ for the linear mappings respectively. These four regions are separated by two hyperplanes, H_1 and H_2. Since f is continuous, the function values have to match at the boundaries of the hyperplanes. For example, going from R_1 to R_2 crossing H_1 means

$$\{x \in H_1 | J_1 x = J_2 x\} \text{ or } \Delta J = 0$$

Since the boundary is of dimension $n-1$, ΔJ has to have rank 1. Then the normal vector defines the whole boundary via the condition $a^T x = 0$. But because the matching at the boundary must hold for each x (thus independent on where the boundary is crossed) and the boundary itself is defined by the normal vector, ΔJ can differ from a^T only by a real constant c, yielding $\Delta J = ca^T$. This amount will be the same for the difference between J_3 and J_4, just because of the independence of crossing the boundary. This process is called a *rank 1 update* of the mapping or a *dyadic update* (see here also our discussion in chapter 1, section 1.6). The relation implies that J_1 changes with a dyadic vector product into J_2 and that it is a valid description throughout the whole boundary. In the literature this relation is known as the *consistent variation property*. As a consequence it is clear that along a closed path Γ (see Fig. 2.1) the summation of the jumps i.e., the changes in the Jacobians, over the two hyperplanes must add up to zero.

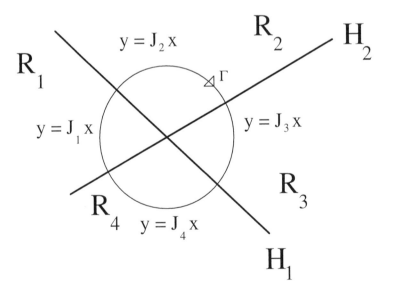

Figure 2.1. The geometry of the domain space

In Fig. 2.1 the domain space is partitioned into four parts. A *degenerated partition* is given by the following definition

Definition 2.1 *An intersection of (n-1) dimensional boundaries in R^n is degenerated if three or more of them meet in an (n-2)dimensional manifold.*

Suppose Fig. 2.1 is valid for R^2. An extra hyperplane, going through the same common point as the already existing planes, will cause the section to degenerate. This because in R^2 two normal vectors can independently be chosen, the third one is dependent on the first two.

In general, there are two ways to define the relation between the input and the output of a function. In *explicit* models for a given input vector the output can be obtained immediately by just substituting the input vector into the function description. For an *implicit* model description this is not true. An algorithm has to be performed by which the output vector can be computed. In the piecewise linear theory, there is a strong relation between *implicit* and *explicit* models. To describe that relation the following definitions will be used.

The absolute value of a vector is given according to

Definition 2.2   *Let $a \in R^n$, then the absolute value $|a|$ is defined as*
$$|a| = (|a_1|, |a_2|, \cdots, |a_n|)^T$$

The *modulus transformation* is needed to transform an implicit description into an explicit description and is given as

Definition 2.3 *Let $z, u, j \in R^n$ and let the n-dimensional vector function $\phi(\cdot)$ be given as $\phi(z)_k = h(z_k)$ where the subscript k denotes the k-th element of a vector and $h(\cdot)$ is a scalar function. For a strictly increasing $h : R_+ \to R_+$ and h(0)=0, the transformation $z \to u, j$ defined by $u = \phi(|z| + z), j = \phi(|z| - z)$ is called the modulus transformation.*

It must be mentioned that the modulus transformation automatically guarantees that $u \geq 0, j \geq 0, u^T j = 0$. Vectors that satisfy the latter three equations are called *complementarity vectors*. If we define $h(\cdot)$ as h(t)=t, corollary 2.1 follows immediately from definition 2.3.

Corollary 2.1 *The modulus transform for h(t)=t is equivalent to the mapping $u, j \to z$ satisfying $|z| = (u + j)/2$ and $z = (u - j)/2$, with $z \in R, u, j \in R_+$.*

As we will see later on, the explicit PL models use the modulus (or absolute-value) operator, while the implicit models use internal complementary state variables. Definition 2.3 defines the relation between the two descriptions. Implicit models use an algorithm to obtain the output vector for a given input vector. The algorithm solves the so-called *Linear Complementary Problem (LCP)* that is defined as [5]

Definition 2.4 *The Linear Complementary problem is defined as to find the complementary vectors u and j for a given q and D such that* $j = Du + q$

In the past several algorithms were developed to solve the LCP and the main important ones will be discussed in chapter 3. However, the possibility to solve the LCP is determined by the class to which matrix D belongs. Here we will define the class P matrix, but in chapter 3 we will return to this issue. One of the matrices in the implicit matrix description must be in this class to guarantee that the mapping given by this description is many-to-one and hence can be transformed into an explicit description.

Definition 2.5 *A matrix D belongs to class P if and only if* $\forall z \in R^p, z \neq 0, \exists k : z_k \cdot (Dz)_k > 0$

Class P is alternatively defined by the property that all principal minors of D are positive [26].

The remaining of this chapter is as follows. First we will discuss the most important and well accepted explicit model descriptions for PL functions after which we will do the same for the implicit models. As discussed in the previous section, the modulus transformation is a strong relation between the implicit and explicit model description. We will use this transformation to rewrite all the models in the same format. Then it becomes possible to compare the models and rank them. The ranking will be done based on the class of continuous functions they can handle. It turns out that the explicit model description so far is not powerful enough and an extension to the model will be treated. Finally we will end up with a discussion on the model description to cover *all* explicit continuous PL functions.

2.2. Explicit PL model descriptions

In this section the most important explicit PL models will be considered. The history of the development of compact explicit piecewise linear models started in 1977 with the paper of Chua and Kang [3]. In this paper they proposed a section-wise PL function and defined for this model the properties. A year later, they came up with the global representation of multi-dimensional PL functions [6]. Since that time many researchers tried to extend this model in the search for the most general model description that can cover all explicit continuous functions, defining a many-to-one mapping. In this section we will treat the most important models nowadays.

2.2.1. The model description of Chua

Consider the one-dimensional function $y = f(x)$ as depicted in Fig. 2.2. The hyperplanes that separate the linear segments in this functions are reduced to points. Hence, each hyperplane only separates two regions in the domain space of dimension R^1. It is not possible that hyperplanes cross each other. In such situation the consistent variation property (see section 2.1) is always satisfied. For this example there are two planes to define:

$$H_1 : x - 1 = 0$$
$$H_2 : x - 2 = 0 \tag{2.1}$$

Suppose that we use the absolute value operator to define on which side of the plane we are and thus to define which segment we finally obtain as function description. Let c_i be constant, representing the amount of contribution of the hyperplane to the function value,

$$c_1 |x - 1|$$
$$c_2 |x - 2| \tag{2.2}$$

Crossing H_1 the Jacobian matrix changes now with the amount

$$\Delta J = J_1 - J_2 = \left(-c_1\right) - \left(c_1\right) = -2c_1 \tag{2.3}$$

as explained in section 2.1. The same holds for the other hyperplane. Because the function is continuous on the boundaries, we obtain the following equations from

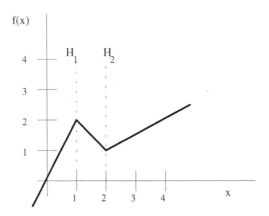

Figure 2.2. An example of the function $f : R^1 \to R^1$

which the values for c_i can be deduced:

$$J_1 - J_2 = 3 = -2c_1 \Rightarrow c_1 = -\frac{3}{2}$$

$$J_2 - J_3 = -\frac{3}{2} = -2c_2 \Rightarrow c_2 = \frac{3}{4}$$

(2.4)

Using (2.2) together with (2.4) and the function definition in $x = \frac{1}{2}$ the function description can then be given by:

$$y = f(x) = \frac{5}{4}x - \frac{3}{2}|x - 1| + \frac{3}{4}|x - 2|$$

(2.5)

The above example shows the general idea behind the model description as proposed by Kang and Chua in 1978. The formal definition of the canonical PL function $f : R^n \rightarrow R^m$ is expressed by

$$f(x) = a + Bx + \sum_{i=1}^{\sigma} c_i |\langle \alpha_i, x \rangle - \beta_i|$$

(2.6)

with $B \in R^{m \times n}, a, c_i \in R^m, \alpha_i \in R^n$ and $\beta_i \in R^1$ for $i \in \{1, \ldots, \sigma\}$. We will refer to this model as *Chua*. Note that also this model can be deduced from an electrical network, see section 1.3.

In this model the domain space R^n is divided into a finite number of polyhedral regions by σ hyperplanes H of dimension n-1. In the model, hyperplane H_i is expressed by

$$\langle \alpha_i, x \rangle - \beta_i = \alpha_i^T x - \beta_i = 0$$

(2.7)

which is also depicted in Fig. 2.3. It is the multi-dimensional extension of (2.1)

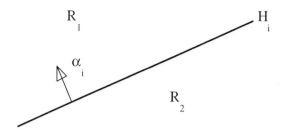

Figure 2.3. Polytope boundary

Hyperplane H_i divides the domain space into two regions, R_1 and R_2. The normal vector of the plane is defined by α_i. When crossing this plane, the Jacobian matrix changes with the amount

$$\Delta J = J_1 - J_2 = \left(-c_i \alpha_i^T\right) - \left(c_i \alpha_i^T\right) = -2c_i \alpha_i^T \qquad (2.8)$$

Notice that this amount is independent of where the hyperplane is crossed.

It must be clear that each one-dimensional function can be realized by (2.6) and that there exist a one-to-one relation between the parameters in (2.6) and the given piecewise linear function. However, in more dimensions, hyperplanes can intersect each other, geometrical constraints might exist, such that not all multi-dimensional functions can be represented by this model description. In [7] it was proven that if the partitioning of the domain space is non degenerated, every PL-function in the strict sense can be given by this model description.

2.2.2. The model description of Güzelis

The problem with the model description *Chua* was that the hyperplanes were planes in the strict sense. This result in the fact that geometrical structures like depicted in Fig. 2.4 could not be modeled in the format (2.6).

To overcome this problem, Güzelis stated that it should be possible to use also hyperplanes that in itself were piecewise linear [8]. To make the model description not unnecessary complex, these PL hyperplanes use already defined hyperplanes in the strict sense to define their boundaries. This is visualized in Fig. 2.4.

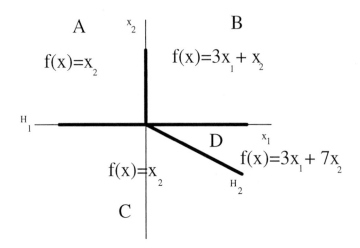

Figure 2.4. Geometric space

Hyperplane H_1 is a boundary in the strict sense. Hyperplane H_2 is a piecewise linear plane that uses H_1 to define its two sections. Such a PL hyperplane H_j can be defined in the model format of *Chua*,

$$\delta_j + \gamma_j^T x + \sum_{i=1}^{\sigma} d_{ji} \left| \alpha_i^T x + \beta_i \right| = 0 \qquad (2.9)$$

These hyperplanes make use of σ boundaries in the strict sense. For the example of Fig. 2.4 it follows from (2.9) that the planes can be defined as $x_2 = 0$ and $x_1 + x_2 - |x_2| = 0$. If one defines τ hyperplanes of the form (2.9) the complete model description according Güzelis is described by

$$f(x) = a + Bx + \sum_{i=1}^{\sigma} b_i \left| \alpha_i^T x + \beta_i \right| + \sum_{j=1}^{\tau} c_j \left| \delta_j + \gamma_j^T x + \sum_{i=1}^{\sigma} d_{ji} \left| \alpha_i^T x + \beta_i \right| \right|$$

$$(2.10)$$

This model is canonical only when the consistent variation property holds. Consider again the situation in Fig. 2.4. For the two adjacent regions D and C with boundary H_2 the Jacobian matrix change must be equal to the matrix change for the two adjacent regions B and A. Thus the change of the matrix must be independent of where the crossing of boundary H_2 takes place. For H_2 this holds,

$$J_D - J_C = (3,7) - (0,1) = \hat{k}_1^1(1,2) \rightarrow \hat{k}_1^1 = 3$$
$$J_B - J_A = (3,1) - (0,1) = \hat{k}_1^2(1,0) \rightarrow \hat{k}_1^2 = 3 \qquad (2.11)$$

The k-factors of the independent hyperplanes are dependent on PL hyperplanes:

$$k_i = \bar{k}_i + \sum_j \hat{k}_j d_{ji} \qquad (2.12)$$

with $\hat{k}_i$ the k-factors belonging to the PL hyperplanes and with $\bar{k}_i$ constant for the crossing. For H_1 one obtains

$$J_A - J_C = (0,1) - (0,1) = k_2^1(0,1) \rightarrow k_2^1 = 0$$
$$J_B - J_D = (3,1) - (3,7) = k_2^2(0,1) \rightarrow k_2^2 = -6 \qquad (2.12)$$

and from this

$$k_2^1 = \bar{k}_2^1 - (3)(-1) \rightarrow k_2^1 = -3$$
$$k_2^2 = \bar{k}_2^2 + (3)(-1) \rightarrow k_2^2 = -3 \qquad (2.13)$$

and thus the consistent variation property holds. It is now easy to obtain the complete function description for Fig. 2.4 that is given by

$$f(x) = \begin{pmatrix} 1.5 & 2.5 \end{pmatrix} \begin{pmatrix} x_1 \\ x_2 \end{pmatrix} + (-1.5)|x_2| + (1.5)|x_1 + x_2 - |x_2|| \qquad (2.14)$$

From (2.10) it can be observed that without the PL hyperplanes, the model description becomes the same as *Chua* and therefore we can conclude that model *Güz* is more general than the one proposed by Chua and Kang.

2.2.3. The model description of Kahlert

To overcome the shortcomings of the previous two model descriptions, Kahlert and Chua came up with a new model (we will refer to this model as model *Kah*). One of the problems with the previous models was that degenerated intersections that do not posses the consistent variation property could not be handled. An example of such a situation is depicted in Fig. 2.5.

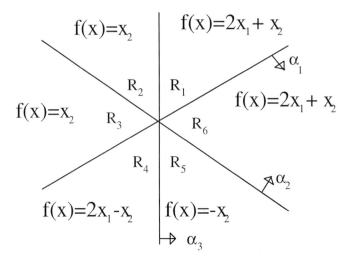

Figure 2.5. An example for the model *Kah*

There are three hyperplanes in a two-dimensional domain space that intersect in one common point. Only two independent normal vectors can be chosen, the other normal vector can be formed by a linear combination from the independent set. Because the hyperplanes are planes in the strict sense, the variation property does not hold for the boundary between regions R_2 and R_3 related to R_5 and R_6.

In [9] the model description is given as

$$f(x) = a + Bx + \sum_{i=1}^{\sigma} c_i \left| \langle \alpha_i, x \rangle - \beta_i \right| + \phi(x) \tag{2.15}$$

with $\phi(x)$ necessary to abolish the violation of the consistent variation property. Assume that $\phi(x) = 0$, then the model becomes equal to *Chua* which possesses the consistent variation property. It was already stated that then each PL-function in with non degenerated domain space can be modeled. For the example in Fig. 2.5 the domain space is degenerated and the function $\phi(x)$ is needed, which is defined as

$$\phi(x) = \sum_{j=1}^{\rho} \sum_{k=3}^{\delta^j} \tilde{c}_{j,jk} \left\{ \left\| a^j_{k,j1} \left(\langle \alpha_{j1}, x \rangle - \beta_{j1} \right) \right| + a^j_{k,j2} \left(\langle \alpha_{j2}, x \rangle - \beta_{j2} \right) \right| - \right.$$
$$\left. \left| a^j_{k,j1} \left(\langle \alpha_{j1}, x \rangle - \beta_{j1} \right) + \left| a^j_{k,j2} \left(\langle \alpha_{j2}, x \rangle - \beta_{j2} \right) \right| \right\| \right\} \tag{2.16}$$

with $\tilde{c}_{j,jk} \in R^m, \alpha_{j1}, \alpha_{j2} \in R^n, \beta_{j1}, \beta_{j2}, a^j_{k,j1}, a^j_{k,j2} \in R^1$.

The first summation of $\phi(x)$ ranges over all degenerate intersections of the partitioning. The inner sum ranges, for every degenerate intersection, over all planes causing the degeneracy. Because such an intersection is degenerate, it is possible to choose a subset of independent normal vectors, where the other normal vectors can be formed by a linear combination from the subset. By substitution a point from every region in $\phi(x)$ it is possible to determine the value of $\phi(x)$. The terms of $\phi(x)$ are added to the consistent variation part of the mapping (i.e. (2.15) with $\phi(x)=0$) and give just enough freedom to remove this consistent variation constraint.

Back to the example of Fig. 2.5. Because there is only one degenerate intersection, the outer sum of (2.16) can be omitted. In this example two normal vectors can be chosen freely, suppose α_1 and α_2. Because the inner sum ranges now over the plane with normal vector α_3 only, this sum can be omitted also. This normal vector can be deduced from $\alpha_3 = a_1\alpha_1 + a_2\alpha_2$ with $a_1 = a_2 = 1$. The combination $a_1\alpha_1 - a_2\alpha_2$ defines the plane $x_2 = 0$. By substituting some points, the function $\phi(x)$ in geometrical sense finally becomes

$$R_1 : 2\,\tilde{c}\,a_2\alpha_2^T x \hat{=} J_\phi^1$$

$$R_2 : -2\,\tilde{c}\,a_1\alpha_1^T x \hat{=} J_\phi^2$$

$$R_3 : 0$$

$$R_4 : 2\,\tilde{c}\,a_2\alpha_2^T x \hat{=} J_\phi^1$$ (2.17)

$$R_5 : -2\,\tilde{c}\,a_1\alpha_1^T x \hat{=} J_\phi^2$$

$$R_6 : 0$$

which is depicted in Fig. 2.6.

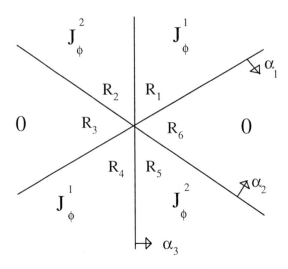

Figure 2.6. Function $\phi(x)$

The function can then be described in the *Kah* format as

$$f(x) = x_1 + x_2$$
$$+ \frac{1}{2}\left\{ \left|x_1 + x_2\right| + \left|x_1 - x_2\right| + \left\| \left|x_1 + x_2\right| + x_1 - x_2 \right| - \left|x_1 + x_2 + \left|x_1 - x_2\right|\right\| \right\}$$

With the model *Kah* every two-dimensional PL-function in the strict sense can be modeled. In [10] Kahlert and Chua proposed a format to extent this model towards higher dimensions. In this paper, they only discuss the possible geometry of the domain space, the model description itself was not discussed.

2.2.4. The model description of Huertas

In 1984, Huertas came up with an explicit model description, which has a slight different format than *Chua*,

$$f(x) = a + Bx + \sum_{i=1}^{\sigma}\left\{ c_i\left|\langle \alpha_i, x\rangle - \beta_i\right| + h_i\,\mathrm{sgn}\big(\langle \alpha_i, x\rangle - \beta_i\big) \right\} \qquad (2.18)$$

with sgn(x) the sign-operator on x [11].

   Although the advantage of this description is that also discontinuous functions can be modeled, there are two serious drawbacks:

- It is not a compact description since it does not take the full advantage of the continuity of a given continuous PL function. Therefore in such situation the model is not canonic.

- There are several uncertainties in the generation of the model (the procedure is not strict).

For these two reasons, the model is not well accepted in the literature and will therefore not be treated further.

2.3.   Implicit PL model descriptions

A few years after the introduction of the first PL model description by Chua, van Bokhoven came up with an implicit model description [2]. This model was derived from a more network theoretical viewpoint as discussed in chapter 1. A small modification was given a few years later. Both models use state variables to define the partition of the domain space.

2.3.1. The model description of van Bokhoven, model 1

In chapter 1, the states of a PL electrical network were represented by the states of the ideal diodes. In such a network each ideal diode has the characteristic

$$u \geq 0, j \geq 0, uj = 0 \qquad\qquad (2.19)$$

and with such a description is was possible to define in which state the network was located. For each state there was exactly one combination of blocking or conducting diodes.

In a more abstract way of thinking, a PL electrical network represents a PL function and with each ideal diode one can define a hyperplane in the domain space. Each hyperplane divides the domain space into two regions. Then the division $j \geq 0$ or $u \geq 0$, but not both positive, defines which half space of domain space is valid. In a one-dimensional situation, this looks like,

$$j = cx + u + g \qquad\qquad (2.20)$$

that defines the plane $cx+g=0$. For $x \geq -\frac{g}{c}$ (right hand side of the hyperplane) u becomes positive and thus j zero. For the left halfspace, the situation is reversed. For each halfspace a linear function description is valid, which must be continuously at the boundary of $cx+g=0$. Suppose that for the left site of the hyperplane ($x \leq -\frac{g}{c}$) the function is defined as

$$y = a_1 x + f_1 \qquad\qquad (2.21a)$$

and for the other side

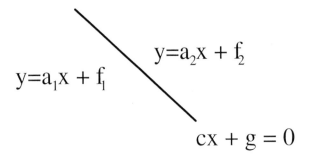

Figure 2.7. One dimensional situation

$$y = a_2 x + f_2 \tag{2.21b}$$

(see also Fig. 2.7) then at the boundary the following relation must hold

$$a_1 x + f_1 = a_2 x + f_2 \tag{2.21c}$$

To assure (2.21c) define the function description as

$$y = ax + bu + f \tag{2.22}$$

For the left site, $u=0$ holds and thus $a = a_1, f = f_1$ from (2.21a). For the other site becomes $j=0$ true and thus from (2.20) $u = -cx - g$. This together with (2.21b) and (2.22) gives

$$y = a_1 x + f_1 + b(-cx - g) = a_2 x + f_2 \tag{2.23}$$

that gives two expressions for b. In fact this means that it is not possible to choose b freely and thus there are restrictions to the linear mappings. These restrictions are exactly the consistent variation restriction for the model. The complete model becomes

$$y = a_1 x + (-\frac{a_2 - a_1}{c})u + f_1 \tag{2.24}$$
$$j = cx + u + g$$

For a given x first the second relation is solved to find whether j or u becomes zero. The first relation gives the correct mapping.

In 1981, van Bokhoven presented a model description, based on the above considerations, for the mapping $f : R^n \rightarrow R^m$ that looks like [2.10]

$$y = Ax + Bu + f$$
$$j = Cx + Du + g \qquad\qquad (2.25)$$
$$u \geq 0, j \geq 0, u^T j = 0$$

with $A \in R^{m \times n}, B \in R^{m \times k}, C \in R^{k \times n}, D \in R^{k \times k}, f \in R^m, g \in R^k$. We will refer to this model as *Bok1*.

In this model description there are k hyperplanes, defined in the second equation. They define in which state the model is and therefore this equation is called the state equation. The vectors u and j are the state vectors. It must become clear from the above considerations that k hyperplanes define 2^k partitions in the domain space, called polytopes. For each polytope there is a linear mapping defined in the first part of (2.25), called the system equation. The last equation in (2.25) defines the restrictions on u and j and theses are related to the so-called Linear Complementary Problem (LCP). We will return to this problem in chapter 3. In section 2.1, it has already been shown that the modulus-operator is closely related to this LCP property.

Comparing (2.24) with (2.25) it turns out that in the first model D is equal to the identity matrix. This means that the hyperplanes were planes in the strict sense. With a full matrix D this is not true anymore and complex geometrical structures can be modeled. It becomes now for instance possible to describe one-to-many mappings.

An example is given in Fig. 2.8 for which the model description can be given as:

$$y = (-1)x + (-1 \quad 1)u + (1)$$

$$\begin{pmatrix} j_1 \\ j_2 \end{pmatrix} = \begin{pmatrix} -1 \\ 1 \end{pmatrix} x + \begin{pmatrix} -1 & 1 \\ 1 & -1 \end{pmatrix} u + \begin{pmatrix} 1 \\ 0 \end{pmatrix} \qquad\qquad (2.26)$$

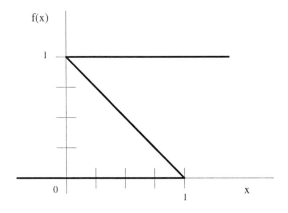

Figure 2.8. One-to-many mapping

For this situation there are two hyperplanes (points), $x=0$ and $x=1$ defining three polytopes:

$$x \leq 1\} \quad f(x) = 0$$

$$\left.\begin{matrix} x \leq 1 \\ x \geq 0 \end{matrix}\right\} f(x) = -x \tag{2.27}$$

$$x \geq 0\} \quad f(x) = 1$$

The fourth polytope cannot be reached, because its constraints on x are $x \leq 0, x \geq 1$.

Although it may not be clear from the description, there is a one-to-one relation between the matrices of the model and a given PL function. However, this is only true for the situation that matrix D is of class P (see definition 2.5). We will return to this issue in section 2.6.

2.3.2. The model description of van Bokhoven, model 2

To gain insight in the geometrical structure of the domain space, it is more convenient to have only hyperplanes in the strict sense. It is then easier to obtain the parameters for the model. But having such hyperplanes only in the domain space will result in a model for a limited class of PL functions. To overcome that problem, van Bokhoven proposed that it should also be possible to define hyperplanes in the image space. This resulted in the following model description to which we will refer as *Bok2*:

$$0 = Iy + Ax + Bu + f$$
$$j = Dy + Cx + Iu + g \tag{2.28}$$
$$u \geq 0, \ j \geq 0, u^T j = 0$$

with $A \in R^{m \times n}, B \in R^{m \times k}, C \in R^{k \times n}, D \in R^{k \times m}, f \in R^m, g \in R^k$ [4].

Due to the unity matrix in front of the state vector u, the description has planes in the strict sense, lying in the domain and image space. Besides the advantage of modeling, this model has several advantages when it is used as concept for simulating. This aspect will be treated in chapter 4.

For the function of Fig. 2.8, the model description can be given as

$$0 = (1)y + (0)x + \begin{pmatrix} -2 & 2 \end{pmatrix}u + (0)$$

$$\begin{pmatrix} j_1 \\ j_2 \end{pmatrix} = \begin{pmatrix} -1 \\ -1 \end{pmatrix}y + \begin{pmatrix} -\frac{1}{2} \\ -\frac{1}{2} \end{pmatrix}x + Iu + \begin{pmatrix} \frac{1}{2} \\ 1 \end{pmatrix} \tag{2.29}$$

with as the hyperplanes:

$$y = -\frac{1}{2}x + \frac{1}{2}$$
$$y = -\frac{1}{2}x + 1$$

(2.30)

that are indeed located in the combined domain – image space.

2.4. Comparison of the models

Above several possible model descriptions were presented to describe the piecewise linear function mapping $f : R^n \rightarrow R^m$. Although the explicit models, *Güz, Kah* and the model of Huertas are extensions to the model *Chua*, it is not clear from the above given outline how the models are related to each other. Which one can cover the largest set of PL functions is a question that is important to answer. The same holds for the implicit models and more important, how are the implicit models related to the explicit models. Those questions will be answered in this section.

To that purpose we will use the modulus transformation, which is treated in section 2.1. This transformation allows us to obtain a direct relation between the modulus operator as used in the explicit model descriptions and the state vectors with there properties as used in the implicit descriptions. For convenience, we will not explicitly repeat to mention the restrictions on u and j, namely $u \geq 0, j \geq 0, u^T j = 0$, however, they still have to be obeyed.

2.4.1. Model Chua

Let us reformulate model *Chua* using matrices, which leads to

$$f(x) = y = a + Bx + C|Ax - b|$$

(2.31)

where the vectors c_i, α_i and the elements β_i in (2.6) are placed as rows in C, A and b respectively.

Define $z = Ax - b$, then the modulus operator (see Corollary 2.1, section 2.1) yields

$$|Ax - b| = \frac{1}{2}(u + j)$$
$$Ax - b = \frac{1}{2}(u - j)$$

(2.32)

Relation (2.31) can be reformulated, by substituting (2.32) into (2.31) and rewriting the second equation of (2.32), to

$$y = (B - CA)x + Cu + (a + Cb)$$
$$j = (-2A)x + Iu + (2b)$$

$$(2.33)$$

that has the same format as *Bok1* with

$$\tilde{A} = B - CA$$

$$\tilde{B} = C$$

$$\tilde{f} = a + Cb$$

$$(2.34)$$

$$\tilde{C} = -2A$$

$$\tilde{D} = I$$

$$\tilde{g} = 2b$$

From (2.33) and (2.34) it can be observed that the model description *Chua* is a sub class of the model description *Bok1*, the *D* matrix in the latter model has now the special form of the identity matrix. Therefore the consistent variation property is always satisfied.

It may also become clear that for a given model of form *Chua*, one can directly obtain the model parameters for *Bok1*, there is a one-to-one relation. Therefore, each implicit model *Bok1* with the unity matrix in front of the state vector *u* in the state equation can be transformed into an explicit form.

Consider the example of Fig. 2.2 with as description (2.5), which is in matrix form written as

$$f(x) = \frac{5}{4}x + \left(-\frac{3}{2} \quad \frac{3}{4} \right) \begin{pmatrix} 1 \\ 1 \end{pmatrix} x - \begin{pmatrix} 1 \\ 2 \end{pmatrix}$$

This can be transformed using (2.34) into

$$y = 2x + \left(-\frac{3}{2} \quad \frac{3}{4} \right) u + (0)$$

$$j = \begin{pmatrix} -2 \\ -2 \end{pmatrix} x + Iu + \begin{pmatrix} 2 \\ 4 \end{pmatrix}$$

with $x=1$ and $x=2$ as hyperplanes which is exactly as defined in (2.1). Each entry in the B-matrix of this model represents the difference in slopes between two adjacent segments. This property could also be observed in for instance (2.24). One can easily prove that this property holds for each one-dimensional function [4].

2.4.2. Model Güz

Model *Güz*, defined in (2.10), can be recast into the form

$$f(x) = y = a + Bx + C|Ax + b| + D|d + Gx + E|Ax + b\|| \qquad (2.35)$$

using the same technique as in reformulating (2.6) into (2.31).
  As before, define for this model

$$|Ax - b| = \frac{1}{2}(u_1 + j_1)$$

$$Ax - b = \frac{1}{2}(u_1 - j_1)$$

$$\qquad (2.36)$$

$$|d + Gx + E|Ax + b\|| = \frac{1}{2}(u_2 + j_2)$$

$$d + Gx + E|Ax + b| = \frac{1}{2}(u_2 - j_2)$$

then by substituting the first and third equation of (2.36) into (2.35) and reformulating the other equations, (2.35) can be transformed into

$$y = (B - CA - DG + DEA)x + (C - DE \quad D)\begin{pmatrix} u_1 \\ u_2 \end{pmatrix} + (a - Cb - Db + DEb)$$

$$\begin{pmatrix} j_1 \\ j_2 \end{pmatrix} = \begin{pmatrix} -2A \\ -2G + 2EA \end{pmatrix} x + \begin{pmatrix} I & 0 \\ -2E & I \end{pmatrix}\begin{pmatrix} u_1 \\ u_2 \end{pmatrix} + \begin{pmatrix} -2b \\ -2d + 2Eb \end{pmatrix}$$

$$\qquad (2.37)$$

that has the same form as *Bok1*. Here the **0** is a matrix with entries equal zero.
  This result means that there is a one-to-one relation between the two model descriptions. In [8] it was stated that there is no systematic method to obtain the coefficients of *Bok1*. This statement does not hold for the class of functions that can be modeled with *Güz* which was already shown in [12].
  In the state equation of (2.37) the first part defines hyperplanes in the strict sense. The second set consists of the piecewise affine hyperplanes that indeed by matrix E depends on the first set of planes, as can also be seen from (2.35).

Obviously, *Güz* belongs to a superset of *Chua* and belongs to the same class if the PL hyperplanes are excluded, i.e. E equals the zero matrix. The model description of *Güz* is less powerful than that of *Bok1* and therefore not each *Bok1* can be transformed into *Güz*. Consider for instance the hysteresis function of (2.26) that has a full matrix D.

Consider again the example of Fig. 2.4 with as function description (2.14). In the form (2.35) this will look like

$$f(x) = (1.5 \quad 2.5)x + (-1.5)\big|(0 \quad 1)x - (0)\big|$$
$$+ (1.5)\big\|(1 \quad 1)x - (0) - (-1)\big|(0 \quad 1)x - (0)\big\|$$

which using (2.37) can be transformed into

$$y = (0 \quad 1)x + (0 \quad 1.5)u + (0)$$
$$j = \begin{pmatrix} 0 & -2 \\ -2 & -4 \end{pmatrix} x + \begin{pmatrix} 1 & 0 \\ 2 & 1 \end{pmatrix} u + \begin{pmatrix} 0 \\ 0 \end{pmatrix}$$

Obviously there is one hyperplane in the strict sense, $x_2 = 0$ and one PL hyperplane $x_1 + 2x_2 = 0$ that will break at $x_2 = 0$. Then the plane changes direction with the amount $u_1 = 2x_2$. Filling this into the second equation will lead to $x_1 = 0$, which is exactly the geometry of Fig. 2.4. From this figure it can be seen that crossing $x_2 = 0$ on the left side of the PL hyperplane will not influence the mapping. This affect is represented by the zero entry in the B-matrix at the place of the $x_2 = 0$ hyperplane.

2.4.3. Model Kah

Also the model description *Kah* can be rewritten into a matrix form, which will look like

$$f(x) = y = a + Bx + C|Ax - b| + C_1|A_1x + b_1| + C_2|A_2x + b_2| +$$
$$D\big\{\big|E(A_1x + b_1) + F(A_2x + b_2)\big| - \big|E(A_1x + b_1) + |F(A_2x + b_2)|\big\|\big\} \qquad (2.38)$$

where the vectors c_i, α_i and the elements β_i in (2.15) are placed as rows i in C, A and b, respectively. The subscript 1 (2) denotes the information belonging to the first (second) independent normal vector α_{j1} (α_{j2}) as mentioned in section 2.2.3.

Using the same technique as before, this equation can be transformed into

$$y = \left(B - CA - C_1A_1 - C_2A_2 + 2DG\right)x + \begin{pmatrix} C & C_1 & C_2 & -D & D & D & -D \end{pmatrix}u$$
$$+ \left(a + Cb + C_1b + C_2b - 2Dg\right)$$

(2.39)

$$j = \begin{pmatrix} -2A \\ -2A_1 \\ -2A_2 \\ -2EA_1 \\ -2FA_2 \\ 2G \\ -2G \end{pmatrix} x + \begin{pmatrix} I & & & & & \\ & I & & & & \\ & & I & & & \\ & & & I & & \\ & & & & I & \\ & & & -2I & & I \\ & & & & -2I & & I \end{pmatrix} u + \begin{pmatrix} 2b \\ 2b_1 \\ 2b_2 \\ 2Eb_1 \\ 2Fb_2 \\ -2g \\ 2g \end{pmatrix}$$

with $G = \left(EA_1 - FA_2\right)$, $g = \left(Eb_1 - Eb_2\right)$.

Consider the state equation of (2.39), where we have 5 sets of hyperplanes in the strict sense and two sets that are dependent on those. The first set is clearly formed out of the single modulus forms of (2.38) where in the B-matrix the corresponding multiplication factors can be found. The PL hyperplanes are dependent on the equalities 4 and 5 in the state equation, and those equalities describe precisely the two independent sets of normal vectors for each degenerated section. This is thus conform the theory as mentioned in section 2.2.3.

It is obvious from (2.39) that model *Kah* is of a larger class than *Chua*. The same holds with respect to model *Güz* due to the fact that there are restriction on the formation of the hyperplanes in this model, as can be seen from relation (2.9). However, we will prove lateron that this is not true for the two-dimensional domain space. In such situation, both models are exactly the same.

The example of Fig. 2.5 in the format *Bok1* can be given by

$$y = \begin{pmatrix} 0 & 2 \end{pmatrix}x + \begin{pmatrix} 0 & \dfrac{1}{2} & \dfrac{1}{2} & -\dfrac{1}{2} & \dfrac{1}{2} & \dfrac{1}{2} & -\dfrac{1}{2} \end{pmatrix}u + (0)$$

$$j = \begin{pmatrix} 0 & 0 \\ -2 & -2 \\ -2 & 2 \\ -2 & -2 \\ -2 & 2 \\ 0 & 4 \\ 0 & -4 \end{pmatrix} x + \begin{pmatrix} 1 & & & & & \\ & 1 & & & & \\ & & 1 & & & \\ & & & 1 & & \\ & & & & 1 & \\ & & & -2 & & 1 \\ & & & & -2 & & 1 \end{pmatrix} u + \begin{pmatrix} 0 \\ 0 \\ 0 \\ 0 \\ 0 \\ 0 \\ 0 \end{pmatrix}$$

in which the planes $x_1 + x_2$ and $x_1 - x_2$ are defined, conform (2.17). From this result it can be seen that the first state equation is redundant and can be removed. Form (2.17) is therefore not canonical.

2.4.4. Model Bok2

Finally, the implicit model *Bok2* will be transformed into *Bok1*. It is obviously that this can easily be achieved by substituting the system equation into the state equation, leading to

$$y = (-A)x + (-B)u + (-f)$$
$$j = (C - DA)x + (I - DB)u + (g - Df)$$

(2.40)

With D equals zero, the planes (in the strict sense) are only in the domain space, the model becomes equal to that of *Chua*. Compared to the model of *Güz*, this model is of a different class that means that there are PL functions which can be represented in *Güz* but not in *Bok2* and vice versa. To show this let the model *Bok2* be given as

$$0 = Iy + Ax + \begin{pmatrix} B_1 & B_2 \end{pmatrix}u + f$$

$$j = \begin{pmatrix} D_1 \\ D_2 \end{pmatrix}y + \begin{pmatrix} C_1 \\ C_2 \end{pmatrix}x + Iu + \begin{pmatrix} g_1 \\ g_2 \end{pmatrix}$$

that can be transformed into *Bok1*:

$$y = -Ax + \begin{pmatrix} -B_1 & -B_2 \end{pmatrix}u - f$$

$$j = \begin{pmatrix} C_1 - D_1 A \\ C_2 - D_{21} A \end{pmatrix}x + \begin{pmatrix} I - D_1 B_1 & -D_1 B_2 \\ -D_2 B_1 & I - D_2 B_2 \end{pmatrix}u + \begin{pmatrix} g_1 - D_1 f \\ g_2 - D_2 f \end{pmatrix}$$

If each PL model of type *Güz* can be transformed into *Bok2*, then the following properties must hold

$$D_2 B_1 = D$$
$$D_1 B_2 = 0$$
$$D_1 B_1 = 0$$
$$D_2 B_2 = 0$$

with $D=-2E$ the matrix from (2.37). It can be shown that the matrix $\begin{pmatrix} D_1 & D_2 \end{pmatrix}^T$ not necessarily contains sufficient freedom to fulfill the properties just given. The opposite is also true. This because non homeomorphic functions (for instance the hysteresis curve) can not be handled by *Güz* which however can be modeled in *Bok2*.

The same statement holds for *Kah*. This description is also not able to handle non homeomorphic functions, but is in itself of a larger class than *Güz*.

2.4.5. Concluding remarks

In the previous section all the model descriptions were rewritten in the format of *Bok1*, which makes it possible to order the models with respect to the class of PL functions they can describe. Model *Chua* belongs to the smallest class, only function with hyperplanes in the strict sense can be modeled. Besides this the function must have the consistent variation property. Extensions were achieved by allowing also PL hyperplanes, resulting into the model *Güz*. To model PL functions which do not posses the consistent variation property, the model of *Kah* was invented. The model *Bok2* differs from the previous mentioned models in the fact that the hyperplanes are also located in the image space. An overall picture is given in the Venn diagram of Fig. 2.9.

The main differences between the implicit models and explicit models are:

- The possibility to model multi-valued functions (i.e. non homeomorphic functions) with the implicit modeled descriptions.
- The direct relation in the implicit models towards the LCP problem. This will give more theoretical insight in the network properties as will be shown in chapter 3 and 6. For instance, it is an important research topic to investigate the number and uniqueness of solutions of a function or in a network.
- The possibility in explicit models to directly obtain the function value for a given input vector.

It must become clear that the model proposed by van Bokhoven, called *Bok1*, turns out to be the most general model description so far known in the literature. Applications with this model will be to advantage in gaining insight in the properties of networks.

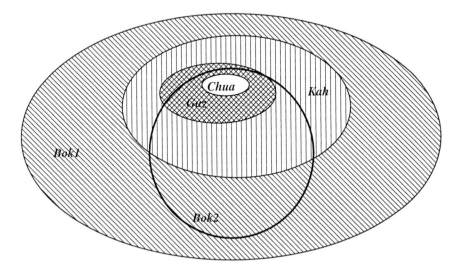

Figure 2.9. Model comparison

2.5.   Extension to the explicit model description

Since the publication of the model of *Chua*, people try to come up with a single explicit model that could cover all explicit PL functions. So far the model of *Kah* is the most powerful one, but there are still geometrical structures possible in dimensions above two that can not be handled by this model. From the model comparison it turns out that the implicit model *Bok1* is of a larger class of *Kah*, which prompted Leenaerts and Kevenaar in 1994 to develop an explicit model description based on this model [13].

2.5.1. The model description

The idea was that in the implicit modeling, the state equation defines the geometry of the space while this in the explicit models is realized by the modulus-operators. The relation between both is the modulus transformation, $|z| = (u + j)/2, z = (u - j)/2$. Furthermore they assumed that D is lower triangular and of class P which implies that all elements on the diagonal are positive. The diagonal elements can be made equal to 1 by appropriate scaling of the vectors u and j. Furthermore, the vector $Cx+g$ is written as $q(x)$ to show its dependence on x. This way, the state equation $j = Cx + Du + g$ can be rewritten as

$$\begin{pmatrix} j_1 \\ \vdots \\ j_k \end{pmatrix} = \begin{pmatrix} 1 & & & & \\ d_{21} & 1 & & \mathbf{0} & \\ d_{31} & d_{32} & & & \\ \vdots & & \ddots & 1 & \\ d_{k1} & \cdots & & d_{k\,k-1} & 1 \end{pmatrix} \begin{pmatrix} u_1 \\ \vdots \\ u_k \end{pmatrix} + \begin{pmatrix} q_1(x) \\ \vdots \\ q_k(x) \end{pmatrix} \qquad (2.41)$$

The first step in the derivation is to find an explicit expression for the vector u that can then be substituted into the system equation $y = Ax + Bu + f$ to give an explicit description for the total mapping $x \rightarrow y$. To find u, the modulus transformation will be used. However, to avoid clutter we will use a special notation for the modulus transformation. From corollary 2.1 we find that

$$|z| = (u + j)/2 \qquad (2.42)$$
$$z = (u - j)/2$$

Next we define the notation

$$\lfloor p \rfloor = (|p| + p)/2 \qquad (2.43)$$

With these definitions it is possible to solve u_1 from the first equation in (2.41) as follows:

$$-q_1(x)/2=(u_1 - j_1)/2=z_1 \qquad (2.44)$$

From (2.42) follows that $u = |z| + z = \lfloor 2z \rfloor$ so that we can write

$$u_1 = \lfloor -q_1(x) \rfloor \qquad (2.45)$$

This is an explicit expression for u_1 which can be used in the second equation of (2.41) to find an explicit form for u_2 as

$$u_2 = \lfloor -q_2(x) - d_{21} \lfloor -q_1(x) \rfloor \rfloor \qquad (2.46a)$$

This back substitution can be repeated to solve every entry in u, leading to

$$u_i = \left\lfloor -q_i(x) - \sum_{k=1}^{i-1} d_{ik} u_k \right\rfloor \qquad (2.46b)$$

The functions u_i thus derived serve as a kind of base functions to transform an implicit description with a lower triangular D into an explicit form. In Fig. 2.10 the first two base functions are depicted. According to the maximum nesting level of the modulus operator, we might call (2.45) a base function of the first order, (2.46a) of the second order and so forth. For every entry in y, the matrix B in the system equation adds the base functions with appropriate weights thus yielding an explicit

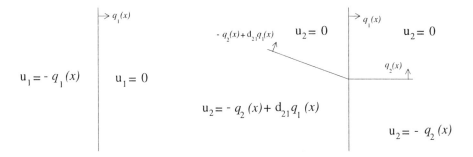

Figure 2.10. First and second order base functions

expression for the mapping $R^n \rightarrow R^m; x \rightarrow y$.

The explicit description just derived is more general than *Kah* because the higher order base functions allow a more complex geometry of the domain space in dimensions higher than R^2. In that sense it is an extension to the known explicit model descriptions. On the other hand it is not the most general explicit form because of the assumption that the matrix D is lower triangular.

2.5.2. Degeneracy of the base functions

From the derivations in the previous section it might appear that the maximum nesting level of the modulus operator is always equal to the size of D. This will however not always be the case. If for instance the entry d_{21} happens to be zero, the second order base function will degenerate into a first order base function thus reducing the nesting level of all higher order base functions. If D is equal to the unit matrix, then all base function will be reduced to first order base functions. However, the effective nesting level follows immediately from the back substitution process.

A question that is far more difficult to answer is the maximum required nesting level. One of the causes that reduce the nesting level was described above. An other possibility is that degeneracy occurs because the functions $q_i(x)$ turn out to be linearly dependent. As an example we assume that $q_2(x)$ depends linearly on $q_1(x)$ so that

$$q_2(x) = \alpha q_1(x) \tag{2.47}$$

The second order base function is then given as

$$u_2 = \left\lfloor -\alpha q_q(x) - d_{21} \left\lfloor -q_1(x) \right\rfloor \right\rfloor \tag{2.48}$$

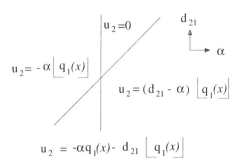

Figure 2.11. Four possible expressions for u_2 depending on α and d_{21}

By analyzing all possibilities of the $\lfloor\ \rfloor$-operator it is possible to find explicit expressions for u_2 depending on the values of α and d_{21} which can be depicted in a picture like Fig. 2.11. Note that all explicit expressions only contain one nesting level. Furthermore it is interesting to notice that in the two small regions either $u_2 \equiv 0$ or $u_2 > 0$ (so $j_2 \equiv 0$) and thus for these values of α and d_{21} the equation for u_2 may be eliminated because it never changes sign and does not contribute to the PL behavior of the mapping. This leads to a reduction of state variables in the implicit representation.

The example suggests that when a new $q_i(x)$ depends linearly on previous $q_i(x)$'s, the degree of nesting does not increase but this was not yet proved. It would imply that the maximum required nesting level is equal to the dimension of the space spanned by the $q_i(x)$'s. For a space of dimension 1 this was proved by Chua [3,6] and for dimension 2 by Kahlert [9].

To illustrate the use of the base functions, some examples will be given.

Example 1, the Chua model

First we take the matrix D equal to the unity matrix I so that we have an expression of the form

$$j = Iu + q(x)$$

It is then easy to see that the base functions u_i are given as

$$u_i = \lfloor -q_i(x) \rfloor = \left| -q_i(x)/2 \right| - q_i(x)/2$$

These are the only base functions that are used in the oldest closed form PL function description introduced by Chua in [3] and which was treated as *Chua*.

Example 2, the Güz model

As a second example the following form is given:

$$j = \begin{pmatrix} 1 & 0 & 0 & 0 \\ 0 & 1 & 0 & 0 \\ a_1 & a_2 & 1 & 0 \\ a_3 & a_4 & 0 & 1 \end{pmatrix} u + q(x)$$

for which we find the following set of base functions

$$u_1 = \lfloor -q_1(x) \rfloor$$

$$u_2 = \lfloor -q_2(x) \rfloor$$
$$u_3 = \lfloor -q_3(x) - a_1 \lfloor -q_1(x) \rfloor - a_2 \lfloor -q_2(x) \rfloor \rfloor$$
$$u_4 = \lfloor -q_4(x) - a_3 \lfloor -q_1(x) \rfloor - a_4 \lfloor -q_2(x) \rfloor \rfloor$$

Note that the functions u_2, u_3 and u_4 are degenerate forms of higher order base functions i.e. due to the special form of the matrix D, some parameters in the base functions had to be chosen equal to 0. This set of base functions is used to build the explicit description *Güz*.

Example 3, the Kah model

As a third example we take the description for *Kah*

$$f(x) = a + Bx + \sum_{i=1}^{\sigma} c_i \left| \langle \alpha_i, x \rangle - \beta_i \right| + \phi(x)$$

with $\phi(x)$ a special summation of expressions of

$$\left\| c_1 \left(\alpha_1^T x + \beta_1 \right) \right| + c_2 \left(\alpha_2^T x + \beta_2 \right) \right\|$$

From this it can be seen that this description can be given using first and second order base functions so that now all the explicit descriptions can be composed from the set of base functions just derived. From this it can also be seen that in R^2 the description of Güzelis covers the same class of mappings as the description of Kahlert.

Example 4, the extended model

From the previous examples again it becomes obvious that the newly derived base functions cover *all* the explicit descriptions given in the literature this far. More specific, the *Güz* model description can be given as a set of higher order (degenerate) base functions. The description of *Kah* is built from a special summation of first and second order base functions and hence the new method also allows to describe any continuous piecewise linear function in R^2. The higher order base functions however allow for a deeper nesting level of the modulus operator and therefore a more complex geometry of the domain space in higher dimensions.

Finally, we demonstrate the use of base functions in an example that cannot be written in one of the explicit forms known up till now. The implicit form is below given where the domain space is chosen equal to R^3.

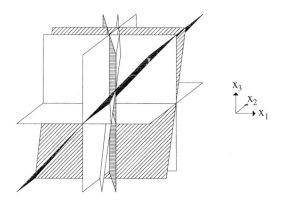

Figure 2.12. The geometry of the domain space

$$y = \begin{pmatrix} -1 & 3 & 5 \\ -2 & -1 & 0 \end{pmatrix} x + \begin{pmatrix} 1 & 3 & 0 \\ 1 & 0 & 4 \end{pmatrix} u$$

$$j = \begin{pmatrix} 1 & 0 & 0 \\ 0 & 1 & 0 \\ 0 & 0 & 1 \end{pmatrix} x + \begin{pmatrix} 1 & 0 & 0 \\ -2 & 1 & 0 \\ 1 & 3 & 1 \end{pmatrix} u$$

In Fig. 2.12 the geometry of the domain space is presented in a picture like in [10]. It can be seen that with only 3 entries in u and j , this space can already be quite complex. The picture is obtained by going through all relevant pivoting steps in the state equation. Because the domain space contains several kinds of degenerate intersections in R^3, it is not possible to represent this function in a form *Kah* or in other explicit PL descriptions presented in the literature.

To write this function in an explicit form, u has to be solved as described before, yielding

$$u_1 = \lfloor -x_1 \rfloor$$
$$u_2 = \lfloor -x_2 + 2\lfloor -x_1 \rfloor \rfloor$$
$$u_3 = \lfloor -x_3 - \lfloor -x_1 \rfloor - 3\lfloor -x_2 + 2\lfloor -x_1 \rfloor \rfloor \rfloor$$

Hence the implicit mapping can be written explicitly where it can be seen that every entry in y is realized by addition of several base functions

$$y = \begin{pmatrix} -1 & 3 & 5 \\ -2 & -1 & 0 \end{pmatrix} \begin{pmatrix} x_1 \\ x_2 \\ x_3 \end{pmatrix} + \begin{pmatrix} 1 & 3 & 0 \\ 1 & 0 & 4 \end{pmatrix} \begin{pmatrix} \lfloor -x_1 \rfloor \\ \lfloor -x_2 + 2\lfloor -x_1 \rfloor\rfloor \\ \lfloor -x_3 - \lfloor -x_1 \rfloor - 3\lfloor -x_2 + 2\lfloor -x_1 \rfloor\rfloor\rfloor \end{pmatrix}$$

2.6. Further extensions to the class of explicit PL model descriptions

In the previous section, a set of base functions was proposed, from which only the first and second order functions were necessary to construct all presented explicit PL model descriptions at that time. Now we will use the same property to further extent the class of explicit PL model descriptions [14]. We will show that using the proposed base functions, all PL mappings with a LCP matrix of class P can be modeled. This restriction is of lesser degree than the 'lower triangular' restriction.

We will start to demonstrate how a full class P matrix can be transformed into a lower triangular matrix. We are then able to transform an implicit PL model description into an explicit description, provided that the related LCP is of class P. We will formulate these necessary steps. Next the first step is treated in more detail, i.e. given a PL mapping find the implicit PL model description. All steps will then rigorous be discussed using an example.

2.6.1. Transformation of a P matrix into an L matrix

In this section we will show that each matrix of class P can be transformed into a lower triangular matrix L. Let us start with the one-dimensional situation, i.e. $P \in R$ and the LCP of (2.25) looks like

$$j = du + q$$
$$u, j \geq 0, uj = 0 \tag{2.49}$$

with d and q scalars. Because $d \in P$, we know that $d > 0$ and hence we easily observe

$$q > 0 \Rightarrow j = q, u = 0$$
$$q \leq 0 \Rightarrow j = 0, u = -q / d \tag{2.50}$$

Using (2.43), the result can be written as an explicit formula

$$j = \lfloor q \rfloor \text{ and } u = \lfloor -q / d \rfloor \tag{2.51}$$

We now extend the dimension to two, yielding the LCP problem

$$j = \begin{pmatrix} d_{11} & d_{12} \\ d_{21} & d_{22} \end{pmatrix} u + \begin{pmatrix} q_1 \\ q_2 \end{pmatrix}$$

$$u, j \geq 0, u^T j = 0$$

(2.52)

Consider the first part of (2.52),

$$j_1 = d_{11} u_1 + r, \qquad r = d_{12} u_2 + q_1$$

(2.53)

From $D \in P$ it follows that $d_{11} > 0$. From (2.49) and (2.51) it follows that

$$j_1 = \lfloor r \rfloor = \lfloor d_{12} u_2 + q_1 \rfloor$$

(2.54)

Suppose $r > 0$ which yields $j_1 = \lfloor r \rfloor = r$ and $u_1 = 0$. From the second part in (2.52) with the assumption that $u_1 = 0$ it follows that

$$u_2 = (j_2 - q_2)/d_{22} = \lfloor -q_2/d_{22} \rfloor$$

(2.55)

and thus

$$j_1 = \lfloor q_1 + d_{12} \lfloor -q_2/d_{22} \rfloor \rfloor$$

(2.56)

A similar procedure will give us also

$$j_2 = \lfloor q_2 + d_{21} \lfloor -q_1/d_{11} \rfloor \rfloor$$

(2.57)

Even if $u_1 \neq 0$, the above results (2.56-2.57) will hold because in this situation $j_1 = 0$ holds, which is also consistent with (2.56) due to (2.43). The above outlined strategy can be repeated for higher orders in a similar way. Via an induction assumption one can prove that in this way the solutions of each LCP of class P can be written explicitly and in a compact way. We will return to this issue in chapter 3, section 3.6. Notice that the result in itself is not remarkable. Because the matrix is of class P there must be a single solution which can explicitly be found by writing down all possible combinations for u and j. In the above outline it is only shown that this solution can be expressed in a closed form. Our ultimate goal is to show that the full matrix D can be transformed into a lower triangular matrix L.

Following the above procedure, one can also find the expressions for the u-vector,

$$u_1 = \left[-\frac{d_{12}}{\Delta} \left| q_2 - \frac{d_{21}}{d_{11}} q_1 \right| + \frac{d_{12}}{\Delta} q_2 - \frac{d_{22}}{\Delta} q_1 \right]$$

$$u_2 = \left[-\frac{d_{21}}{\Delta} \left| q_1 - \frac{d_{12}}{d_{22}} q_2 \right| + \frac{d_{21}}{\Delta} q_1 - \frac{d_{11}}{\Delta} q_2 \right] \tag{2.57}$$

$$\Delta = d_{11}d_{22} - d_{12}d_{21}$$

To rewrite (2.52) we need to define two additional equations. This means that the matrix L will have a higher dimension than D. This is not surprising, because to solve an LCP will demand an exponential amount of work. Solving L will demand polynomial amount of work and therefore if the transformation would also cost a polynomial amount of work, we could solve an LCP also polynomial wich is a contradiction. The exponential amount of work is in the transformation from D into L. The system of (2.52) can be rewritten into:

$$\begin{pmatrix} j_{a1} \\ j_{a2} \\ j_1 \\ j_2 \end{pmatrix} = \begin{pmatrix} 1 & 0 & 0 & 0 \\ 0 & 1 & 0 & 0 \\ \frac{d_{12}}{\Delta} & 0 & 1 & 0 \\ 0 & \frac{d_{21}}{\Delta} & 0 & 1 \end{pmatrix} \begin{pmatrix} u_{a1} \\ u_{a2} \\ u_1 \\ u_2 \end{pmatrix} + \begin{pmatrix} -q_2 + \frac{d_{21}}{d_{11}} q_1 \\ -q_1 + \frac{d_{12}}{d_{22}} q_2 \\ \frac{d_{22}}{\Delta} q_1 - \frac{d_{12}}{\Delta} q_2 \\ \frac{d_{11}}{\Delta} q_2 - \frac{d_{21}}{\Delta} q_1 \end{pmatrix} \tag{2.58}$$

where the first two rows are the additional equations to define the inner $\lfloor \bullet \rfloor$-terms of (2.57). In this way each LCP of class P can be transformed into an LCP with a lower triangular matrix form. Notice that (2.58) has a similar form of (2.39), because the dimension of the example is two. For higher dimensional problems more complex relations are involved and more or all the entries in the lower part of the matrix will be unequal zero. The base functions are here also used to express the explicit solutions of (2.57).

2.6.2. The explicit model description

By now we know that a full matrix of class P can be transformed into a lower triangular matrix. Further we have a closed compact form to write the solutions for the complementary vectors u and j. With this knowledge we can write down the three steps to come up with an explicit model description in terms of absolute-value operators for any given PL mapping $f : R^n \rightarrow R^m ; x \rightarrow f(x)$.

Step 1: There is a one-to-one relation between a given PL mapping and the parameters of the implicit PL model description of van Bokhoven, provided that the mapping is related to an LCP of class P [15,16]. In the next section, we will demonstrate how the model parameters can be obtained. The mapping can be modeled according to

$$\begin{cases} y = Ax + Bu + f \\ j = Cx + Du + g \\ u^T j = 0, u \geq 0, j \geq 0 \end{cases}$$

Step 2: Using the outlined techniques in section IV, the implicit model description can be transformed into an explicit model description,

$$\begin{cases} y = Ax + \bar{B}\bar{u} + f \\ \bar{j} = \bar{C}x + L\bar{u} + \bar{g} \\ \bar{u}^T \bar{j} = 0, \bar{u} \geq 0, \bar{j} \geq 0 \end{cases}$$

where the bar denotes that the dimensions of the vectors and matrices differ from the original ones. This model is already explicit because one can solve explicitly the second equation due to the lower diagonal form of L. Therefore one can obtain explicitly the output vector for a given input vector.

Step 3: Finally the model can be written using the absolute-value operator using corollary 2.1. The resulting model description will have nested absolute-value operators where the level of nesting can be of any order.

Before we show how step one can be performed we have to make some remarks:

1. Although it is difficult to write down explicitly, there is a one-to-one relation between the mapping parameters and the final model parameters for the model in step 2. This because there is a direct relation between the result of step 1 and step 2 and similar between step 2 and step 3.
2. The base functions turn out also to be valid to model mappings with an LCP of class P.
3. For the one- and two-dimensional situations, the solutions are consistent with already presented explicit models in the literature.
4. The complete procedure could be made automatic, because all steps can be performed in an algorithmic way.

2.6.3. Obtaining the model parameters

Up to now we have demonstrated that for a given continuous piecewise linear function it is possible to come up with an explicit piecewise linear model description under the condition that the LCP matrix is of class P. However, the first step is to come to the implicit PL model description of the given mapping. In this section we will explain how this model can be derived [15,16].

Assume that a general PL mapping is given as N segments σ_i $i = 1,...,N$ and for each segment

- The linear mapping $y = A_i x + f_i$, $A_i \in R^{m \times n}, f_i \in R^m$

- The boundaries of the segment described by $K_i x + s_i \geq 0$, $K_i \in R^{k \times n}, s \in R^k$

Let us consider the situation that two boundary planes intersect each other, thus forming four segments. On their common intersection the two planes are allowed to cause a a bend in their mutual direction. For each segment we have the data as described above. For clarity see Fig. 2.13. The arrows indicate the direction of the normal vectors.

Due to the continuity of the mapping, two segments with a boundary plane in common, given by $cx+g$, must obey at the boundary the following constraints:

$$A_i - A_j = b \cdot c^T$$
$$f_i - f_j = b \cdot g$$

(2.59)

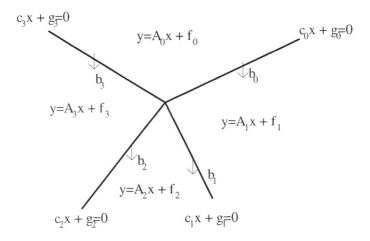

Figure 2.13. Mapping with two boundary planes

that means that for given mappings and boundary planes the only freedom we have is to choose vector b. For the situation as depicted in Fig. 2.13 we obtain

$$\begin{pmatrix} c_2^T \\ g_2 \end{pmatrix} = \lambda_{11} \begin{pmatrix} c_0^T \\ g_0 \end{pmatrix} + \lambda_{12} \begin{pmatrix} c_1^T \\ g_1 \end{pmatrix}$$

$$\begin{pmatrix} c_3^T \\ g_3 \end{pmatrix} = \lambda_{21} \begin{pmatrix} c_0^T \\ g_0 \end{pmatrix} + \lambda_{22} \begin{pmatrix} c_1^T \\ g_1 \end{pmatrix} \qquad (2.60)$$

$$b_0 \cdot c_0^T - b_1 \cdot c_1^T - b_2 \cdot c_2^T + b_3 \cdot c_3^T = 0$$

$$b_0 \cdot g_0 - b_1 \cdot g_1 - b_2 \cdot g_2 + b_3 \cdot g_3 = 0$$

when we follow the path $A_0 \rightarrow A_1 \rightarrow A_2 \rightarrow A_3 \rightarrow A_0$. Filling in the first part of (2.60) into the second part, leads to

$$b_2 = \frac{\lambda_{22}}{\Delta} b_0 + \frac{\lambda_{21}}{\Delta} b_1$$

$$b_2 = \frac{\lambda_{12}}{\Delta} b_0 + \frac{\lambda_{11}}{\Delta} b_1 \qquad (2.61)$$

where $\Delta = \lambda_{11}\lambda_{22} - \lambda_{12}\lambda_{21}$. Because we want to control the mapping and we are not interested in the magnitude of the normal vectors of the boundary planes, we restrict ourselves to $\lambda_{11} = \lambda_{22} = 1$. For a given mapping, choosing b_0 and b_1 according to (2.59) all the information is provided, leading to the model

$$y = A_0 x + \begin{pmatrix} b_0 & b_1 \end{pmatrix} u + f_0$$

$$j = \begin{pmatrix} c_0 \\ c_1 \end{pmatrix} x + \begin{pmatrix} 1 & -\lambda_{12} \\ -\lambda_{21} & 1 \end{pmatrix} u + \begin{pmatrix} g_0 \\ g_1 \end{pmatrix} \qquad (2.62)$$

where the vectors u and j define which segment is valid.

When more boundary planes are involved, the following data should be provided for the pth plane:

- c_p and g_p to define the boundary plane in the segment

- b_p to define the update on the matrix A on the other side of the boundary plane

- $\lambda_{i,p+1}, \lambda_{p+1,i}$ for each $i, 1 \le i \le p$ to define the mutual interaction of this plane with all other boundary planes.

Notice that when $\lambda_{12} = \lambda_{21} = 0$ the boundary planes are strict planes in the sense that after their intersection the planes continue without changing direction. The general expression is therefore given by

$$
\left\{
\begin{array}{l}
y = Ax + Bu + f \\
j = Cx + Du + g \\
u^T j = 0, u \geq 0, j \geq 0
\end{array}
\right.
\tag{2.63}
$$

In this way, if the PL mapping is consistent to an LCP of class P, the mapping can always be transformed into the implicit PL model description of (2.63).

2.6.4 An example

We will demonstrate the method on an example and to this purpose we will examine the three steps as discussed in section 2.6.2. Suppose the following continuous piecewise linear function is given, where we use the definitions from section V ($y=Ax+f$, $Cx+g\geq0$).

$$
R_1: \quad A_1 = \begin{pmatrix} 1 & 0 & -1 \end{pmatrix}, \quad f_1 = \begin{pmatrix} 0 \end{pmatrix}
$$

$$
C_1 = \begin{pmatrix} 1 & 0 & -1 \\ 2 & 1 & -1 \\ -1 & 2 & -1 \end{pmatrix}, \quad g_1 = \begin{pmatrix} 3 \\ -1 \\ 1 \end{pmatrix}
\tag{2.64a}
$$

$$
R_2: \quad A_2 = \begin{pmatrix} 0 & 0 & 0 \end{pmatrix}, \quad f_2 = \begin{pmatrix} -3 \end{pmatrix}
$$

$$
C_2 = \begin{pmatrix} -\frac{1}{2} & 0 & \frac{1}{2} \\ \frac{3}{2} & 1 & -\frac{1}{2} \\ -\frac{3}{2} & 2 & -\frac{1}{2} \end{pmatrix}, \quad g_2 = \begin{pmatrix} -\frac{3}{2} \\ -\frac{1}{2} \\ -3 \end{pmatrix}
\tag{2.64b}
$$

$$
R_3: \quad A_3 = \begin{pmatrix} 1 & 0 & -1 \end{pmatrix}, \quad f_3 = \begin{pmatrix} 0 \end{pmatrix}
$$

$$
C_3 = \begin{pmatrix} \frac{4}{3} & -\frac{2}{3} & -\frac{2}{3} \\ \frac{4}{3} & \frac{7}{3} & -\frac{5}{3} \\ \frac{1}{3} & -\frac{2}{3} & \frac{1}{3} \end{pmatrix}, \quad g_3 = \begin{pmatrix} \frac{8}{3} \\ -\frac{1}{3} \\ -\frac{1}{3} \end{pmatrix}
\tag{2.64c}
$$

$$
R_4: \quad A_4 = \begin{pmatrix} \frac{3}{2} & 1 & -\frac{3}{2} \end{pmatrix}, \quad f_4 = \begin{pmatrix} -\frac{7}{2} \end{pmatrix}
$$

$$
C_4 = \begin{pmatrix} -\frac{1}{2} & 0 & \frac{1}{2} \\ -\frac{3}{2} & -1 & \frac{1}{2} \\ -3 & 1 & 0 \end{pmatrix}, \quad g_4 = \begin{pmatrix} -3 \\ \frac{1}{2} \\ -\frac{5}{2} \end{pmatrix}
\tag{2.64d}
$$

R_5: $\quad A_5 = \left(-\frac{3}{5} \quad \frac{17}{10} \quad -\frac{1}{2}\right), \qquad f_5 = \left(-\frac{21}{4}\right)$

$$C_5 = \begin{pmatrix} -\frac{4}{5} & \frac{1}{10} & \frac{1}{2} \\ 0 & -\frac{3}{2} & \frac{1}{2} \\ \frac{3}{5} & -\frac{1}{5} & 0 \end{pmatrix}, \quad g_5 = \begin{pmatrix} -\frac{13}{4} \\ -\frac{5}{4} \\ \frac{1}{2} \end{pmatrix} \qquad (2.64\text{e})$$

R_6: $\quad A_6 = \begin{pmatrix} 3 & 1 & -2 \end{pmatrix}, \qquad f_6 = (-1)$

$$C_6 = \begin{pmatrix} 1 & 0 & -1 \\ -2 & -1 & 1 \\ -3 & 1 & 0 \end{pmatrix}, \quad g_6 = \begin{pmatrix} 3 \\ 1 \\ 2 \end{pmatrix} \qquad (2.64\text{f})$$

R_7: $\quad A_7 = \left(-\frac{3}{5} \quad \frac{4}{5} \quad -\frac{1}{5}\right), \qquad f_7 = \left(-\frac{21}{5}\right)$

$$C_7 = \begin{pmatrix} -\frac{4}{5} & \frac{2}{5} & \frac{2}{5} \\ 0 & 3 & -1 \\ \frac{3}{5} & -\frac{4}{5} & \frac{1}{5} \end{pmatrix}, \quad g_7 = \begin{pmatrix} -\frac{21}{10} \\ -\frac{7}{2} \\ \frac{6}{5} \end{pmatrix} \qquad (2.64\text{g})$$

R_8: $\quad A_8 = \left(\frac{9}{5} \quad \frac{7}{5} \quad -2\right), \qquad f_8 = \left(-\frac{1}{5}\right)$

$$C_8 = \begin{pmatrix} \frac{8}{5} & -\frac{1}{5} & -1 \\ -\frac{4}{5} & -\frac{7}{5} & 1 \\ \frac{3}{5} & -\frac{4}{5} & 0 \end{pmatrix}, \quad g_8 = \begin{pmatrix} \frac{13}{5} \\ \frac{1}{5} \\ -\frac{2}{5} \end{pmatrix} \qquad (2.64\text{h})$$

Step 1

First we have to find the implicit PL model description. From section 2.6.3 we know that the independent information of two adjacent mappings is in the normal vector, describing their common boundary. This means that for adjacent mappings A_i and A_j their C-matrices must be equal or only differ a constant factor, the λ_{ii}-factor. The normal vector, defining the boundary, can then be found in the C_j-matrix as being the row (suppose k) with opposite sign compared to the same row in C_i,

$$n_{kj} = -\frac{n_{ki}}{\lambda_{ii}} \qquad (2.65)$$

where for n the first index denotes the row number and the second index the mapping.

Considering (2.64) this leads to

$$
\begin{aligned}
A_1 \rightarrow A_2 && \lambda_{11} = 2 \\
A_1 \rightarrow A_6 && \lambda_{22} = 1 \\
A_1 \rightarrow A_3 && \lambda_{33} = 3
\end{aligned}
\tag{2.66}
$$

It follows from section 2.6.3 that only the mappings from (2.66) are important. The other mappings can be deduced from the provided information. Using the normal vectors of the boundaries for the regions 1, 2, 3 and 6 we can obtain λ_{ij}. Suppose that we consider region 1 as the primal region. This means that in the PL model description the A, C, f and g matrices/vectors are equal to those for region 1. First we are investigating the D-matrix. Consider the mappings for region 1 and 2. The rows in their C-matrices define the normal vectors and these are dependent of each other. From section 2.6.3 we have

$$
n_{12} = -\frac{n_{11}}{\lambda_{11}}
$$

$$
n_{22} = n_{21} - \frac{\lambda_{12}}{\lambda_{11}} n_{11}
\tag{2.67}
$$

$$
n_{32} = n_{31} - \frac{\lambda_{13}}{\lambda_{11}} n_{11}
$$

where the first part was already obtained via (2.65). Equation (2.67) gives us $\lambda_{12} = \lambda_{13} = 1$. Similar expressions lead to $\lambda_{21} = 0, \lambda_{32} = 1, \lambda_{13} = 1$ and $\lambda_{23} = -2$. The remaining matrix to investigate is the B-matrix. Using (2.59) we obtain the B-matrix entries, $b_1 = 2, b_2 = -1$ and $b_3 = 0$ which finally gives us the complete PL model description

$$
y = \begin{pmatrix} 1 & 0 & -1 \end{pmatrix} x + \begin{pmatrix} 2 & -1 & 0 \end{pmatrix} u + \begin{pmatrix} 0 \end{pmatrix}
$$

$$
\begin{pmatrix} j_1 \\ j_2 \\ j_3 \end{pmatrix} = \begin{pmatrix} 1 & 0 & -1 \\ 2 & 1 & -1 \\ -1 & 2 & -1 \end{pmatrix} x + \begin{pmatrix} 2 & 0 & -1 \\ 1 & 1 & -2 \\ 1 & 1 & 3 \end{pmatrix} \begin{pmatrix} u_1 \\ u_2 \\ u_3 \end{pmatrix} + \begin{pmatrix} 3 \\ -1 \\ 1 \end{pmatrix}
\tag{2.68}
$$

Step 2

In this step we are intending to transform (2.68) into an equivalent expression with a lower triangular matrix for the D-matrix. This is only possible if $D \in P$ which is obviously the situation for this example. From definition 2.5 it follows that all principal minors are positive, $D_{11} = 5$, $D_{22} = 5$, $D_{33} = 2$, $D_{11,22} = 3$, $D_{11,33} = 1$ and $D_{22,33} = 2$. For simplicity reasons, define

$$
\begin{aligned}
q_1 &= \begin{pmatrix} 1 & 0 & -1 \end{pmatrix} x + \begin{pmatrix} 3 \end{pmatrix} \\
q_2 &= \begin{pmatrix} 2 & 1 & -1 \end{pmatrix} x + \begin{pmatrix} -1 \end{pmatrix} \\
q_3 &= \begin{pmatrix} -1 & 2 & -1 \end{pmatrix} x + \begin{pmatrix} 1 \end{pmatrix}
\end{aligned}
\tag{2.69}
$$

The second part of (2.68) can then also be written as

$$
\begin{pmatrix} u_1 \\ u_2 \\ u_3 \end{pmatrix} = \frac{1}{10} \begin{pmatrix} 5 & 1 & -1 \\ -5 & 5 & 5 \\ 0 & -2 & 2 \end{pmatrix} \begin{pmatrix} j_1 \\ j_2 \\ j_3 \end{pmatrix} + \begin{pmatrix} q_1' \\ q_2' \\ q_3' \end{pmatrix}
\tag{2.70}
$$

with

$$
\begin{aligned}
q_1' &= -\tfrac{1}{2} q_1 - \tfrac{1}{10} q_2 + \tfrac{1}{10} q_3 \\
q_2' &= \tfrac{1}{2} q_1 - \tfrac{1}{2} q_2 - \tfrac{1}{2} q_3 \\
q_3' &= \tfrac{1}{5} q_2 - \tfrac{1}{5} q_3
\end{aligned}
$$

According to the previous sections, u_1 can now be expressed as

$$
u_1 = \left\lfloor \tfrac{1}{10} j_2 - \tfrac{1}{10} j_3 + q_1' \right\rfloor
\tag{2.71}
$$

under the restriction that $j_1=0$. Now we have to examine the situation for j_2 and j_3 under the same restriction. From (2.70) and taking the inverse we have

$$
\begin{pmatrix} j_2 \\ j_3 \end{pmatrix} = \begin{pmatrix} 2 & -5 \\ 2 & 5 \end{pmatrix} \begin{pmatrix} u_2 \\ u_3 \end{pmatrix} + \begin{pmatrix} -2q_2' + 5q_3' \\ -2q_2' - 5q_3' \end{pmatrix}
\tag{2.72}
$$

from which we obtain $j_2 = \left\lfloor -5u_3 - 2q_2' + 5q_3' \right\rfloor$ under the restriction $u_2=0$. Using this restriction and (2.72) we have $u_3 = \left\lfloor \tfrac{2}{5} q_2' + q_3' \right\rfloor$. Combining these two expressions leads to

$$
j_2 = \left\lfloor -5 \left\lfloor \tfrac{2}{5} q_2' + q_3' \right\rfloor - 2q_2' + 5q_3' \right\rfloor .
$$

Following a similar procedure for j_3 gives us the complete expression for u_1

$$u_1 = \left\lfloor \tfrac{1}{10} \left\lfloor -5 \left\lfloor \tfrac{2}{5} q_2' + q_3' \right\rfloor - 2q_2' + 5q_3' \right\rfloor - \tfrac{1}{10} \left\lfloor 2 \left\lfloor q_2' - \tfrac{5}{2} q_3' \right\rfloor - 2q_2' - 5q_3' \right\rfloor + q_1' \right\rfloor$$

$$(2.73a)$$

This is an expression with a three level deep absolute-sign operator. The same procedure gives also

$$u_2 = \left\lfloor -\tfrac{1}{2} \left\lfloor \lfloor q_3' \rfloor - 2q_1' - q_3' \right\rfloor + \tfrac{1}{2} \left\lfloor -5q_3' \right\rfloor + q_2' \right\rfloor$$

$$u_3 = \left\lfloor -\tfrac{1}{5} \left\lfloor -5q_1' + q_2' \right\rfloor + q_3' \right\rfloor$$

$$(2.73b)$$

and using (2.70) this yields

$$u_1 = \left\lfloor 0.1 \left\lfloor -5 \left\lfloor 0.2q_1 - 0.4q_3 \right\rfloor - q_1 + 2q_3 \right\rfloor - 0.1 \left\lfloor 2 \left\lfloor 0.5q_1 - q_2 \right\rfloor - q_1 + 2q_3 \right\rfloor - 0.5q_1 - 0.1q_2 + 0.1q_3 \right\rfloor$$

$$u_2 = \left\lfloor -0.5 \left\lfloor \lfloor 0.2q_2 - 0.2q_3 \rfloor - q_1 + 0.8q_2 + 1.2q_3 \right\rfloor + 0.5 \left\lfloor -q_2 + q_3 \right\rfloor + 0.5q_1 - 0.5q_2 - 0.5q_3 \right\rfloor$$

$$u_3 = \left\lfloor -0.2 \left\lfloor 3q_1 - q_3 \right\rfloor + 0.2q_2 - 0.2q_3 \right\rfloor$$

$$(2.74)$$

The first part of (2.74) can be reformulated into the standard format using four additional complementary equations:

$$j_{a1} = u_{a1} + (0.4q_3 - 0.2q_1) \qquad\qquad \Rightarrow u_{a1} = \left\lfloor -(0.4q_3 - 0.2q_1) \right\rfloor$$

$$j_{a2} = 5u_{a1} + u_{a2} + (q_1 - 2.2q_3) \qquad \Rightarrow u_{a2} = \left\lfloor -5u_{a1} - (q_1 - 2.2q_3) \right\rfloor$$

$$j_{a3} = u_{a3} + (q_2 - 0.5q_1) \qquad\qquad \Rightarrow u_{a3} = \left\lfloor -(q_2 - 0.5q_1) \right\rfloor$$

$$j_{a4} = -2u_{a4} + u_{a1} + (q_1 - 2q_3) \qquad \Rightarrow u_{a4} = \left\lfloor 2u_{a3} - (q_1 - 2q_3) \right\rfloor$$

$$(2.75)$$

leading to

$$j_1 = -0.1u_{a2} + 0.1u_{a4} + u_1 + 0.5q_1 + 0.1q_2 - 0.1q_3 \qquad (2.76)$$

From (2.75) and (2.76) one can already observe the lower triangular format of the LCP matrix. Following this procedure, one finally gets the transformed PL description with as LCP matrix

$$L = \begin{pmatrix} 1 & 0 & 0 & 0 & 0 & 0 & 0 & 0 & 0 & 0 & 0 \\ 5 & 1 & 0 & 0 & 0 & 0 & 0 & 0 & 0 & 0 & 0 \\ 0 & 0 & 1 & 0 & 0 & 0 & 0 & 0 & 0 & 0 & 0 \\ 0 & 0 & -2 & 1 & 0 & 0 & 0 & 0 & 0 & 0 & 0 \\ 0 & 0 & 0 & 0 & 1 & 0 & 0 & 0 & 0 & 0 & 0 \\ 0 & 0 & 0 & 0 & -1 & 1 & 0 & 0 & 0 & 0 & 0 \\ 0 & 0 & 0 & 0 & 0 & 0 & 1 & 0 & 0 & 0 & 0 \\ 0 & 0 & 0 & 0 & 0 & 0 & 0 & 1 & 0 & 0 & 0 \\ 0 & -0.1 & 0 & 0.1 & 0 & 0 & 0 & 0 & 1 & 0 & 0 \\ 0 & 0 & 0 & 0 & 0 & 0.5 & -0.5 & 0 & 0 & 1 & 0 \\ 0 & 0 & 0 & 0 & 0 & 0 & 0 & 0.2 & 0 & 0 & 1 \end{pmatrix} \qquad (2.77)$$

Notice that the format of the matrix differs from the formats of the matrices of Kahlert and Güzelis. This means that with their model descriptions it is not possible to model the mapping discussed in this example.

Step 3
To perform step 3, only the results of (2.74) are important. Using (2.43), u_1 can be expressed as

$$u_1 = 0.5|p| + 0.5p$$
$$p = 0.05\left|-2.5\left|0.2q_1 - 0.4q_2\right| - 1.5q_1 + 3q_2\right| - 0.05\left|\left|0.5q_1 - q_2\right| - 0.5q_1 - q_2 + 2q_3\right|$$
$$- 0.125\left|0.2q_1 - 0.4q_3\right| - 0.05\left|0.5q_1 - q_2\right| - 0.55q_1 - 0.05q_2 + 0.15q_3$$

$$(2.78)$$

Similar expressions can be obtained for u_2 and u_3. The final expression for $y=f(x)$ with the absolute-sign operator can be obtained using the expression $y = \begin{pmatrix} 1 & 0 & -1 \end{pmatrix}x + 2u_1 - u_2$ which gives us

$$y = \begin{pmatrix} 0.5375 & 1.2625 & -0.925 \end{pmatrix}x - 3.625$$

$$+ \left| 0.05 \left| -2.5 \left| (0.6 \quad -0.8 \quad 0.2)x + 0.2 \right| + (-4.5 \quad 6 \quad -1.5)x - 1.5 \right| \right.$$

$$\left. -0.05 \left| (-1.5 \quad -1 \quad 0.5)x + 2.5 \right| + (-4.5 \quad 3 \quad -0.5)x + 1.5 \right|$$

$$-0.125 \left| (0.6 \quad -0.8 \quad 0.2)x + 0.2 \right| - 0.05 \left| (-1.5 \quad -1 \quad 0.5)x + 2.5 \right|$$

$$+ (-0.8 \quad 0.25 \quad 0.45)x - 0.95 \Big|$$

$$-0.5 \left| -0.25 \left| 0.5 \left| (0.6 \quad -0.2 \quad 0)x - 0.2 \right| + (-0.3 \quad 3.1 \quad -1)x - 2.8 \right| \right.$$

$$\left. -0.125 \left| (0.6 \quad -0.2 \quad 0)x - 0.2 \right| + 0.25 \left| (-3 \quad 1 \quad 0)x + 2 \right| \right.$$

$$+ (-0.675 \quad -2.025 \quad 0.75)x + 2.7 \Big|$$

$$+ 0.05 \left| -2.5 \left| (0.6 \quad -0.8 \quad 0.2)x + 0.2 \right| + (-4.5 \quad 6 \quad -1.5)x - 1.5 \right|$$

$$-0.05 \left| (-1.5 \quad -1 \quad 0.5)x + 2.5 \right| + (-4.5 \quad 3 \quad -0.5)x + 1.5 \Big|$$

$$-0.125 \left| (0.6 \quad -0.8 \quad 0.2)x + 0.2 \right| - 0.05 \left| (-1.5 \quad -1 \quad 0.5)x + 2.5 \right|$$

$$+ 0.125 \left| 0.5 \left| (0.6 \quad -0.2 \quad 0)x - 0.2 \right| + (-0.3 \quad 3.1 \quad -1)x - 2.8 \right|$$

$$+ 0.0625 \left| (0.6 \quad -0.2 \quad 0)x - 0.2 \right| - 0.125 \left| (-3 \quad 1 \quad 0)x + 2 \right|$$

$$(2.79)$$

Equation (2.79) is the result of the three steps and the explicit PL model description of the mapping of (2.64). The nesting of the absolute-value operators is three levels deep and hence the model can not be represented by model descriptions previously presented in the literature, i.e. the models of Kahlert and Güzelis. This concludes our example.

2.7. Theorem of Kolmogorov

In the previous section, the extensions to the explicit model description were obtained by assuming the state matrix D of class P. Starting from the implicit model descriptions and some theorems from the neural network theory one can make the connection between modeling and the theorem of Kolmogorov. This leads to an other viewpoint to the format of the model that covers all explicit functions.

2.7.1. The theorem of Kolmogorov

At the turn of the century, during the Paris International Congress of mathematicians, the prominent German mathematician David Hilbert announced a list of difficult problems for the 20th century mathematicians to solve. From this list the 13th problem became famous because of the expected difficulty to solve it. In this problem Hilbert conjectured that there are analytic functions of three variables which cannot be represented as a finite superposition of continuous functions of only two arguments. In a series of papers in the mid to late 1950's Kolmogorov and Arnol'd, both Soviet mathematicians, tried in successive papers to solve this thirteenth problem of Hilbert. Kolmogorov finally succeeded and his result was an astounding theorem concerning the representation of arbitrary continuous functions, mapping an n-dimensional cube onto real numbers, in terms of one dimensional functions [17].

Let I=[0,1] denote the closed unit interval, $I^n = [0,1]^n (n \geq 2)$ the Cartesian product of I. In 1957 Andrei Nikolaevic Kolmogorov proved that any continuous function $f(x_1, \cdots, x_n)$ of several variables defined on $I^n (n \geq 2)$ could be represented in the form

$$f(x) = \sum_{j=1}^{2n+1} \chi_j (\sum_{i=1}^{n} \psi_{ij}(x_i))$$

(2.80)

where χ_j, ψ_{ij} are continuous functions of one variable and ψ_{ij} are monotone functions which are not dependent on f.

In 1965 Sprecher refined the above theorem and obtained the following [18]. For each integer n≥2 there exists a real, monotone increasing function ψ dependent on n and having the following property: for each pre assigned number $\delta > 0$ there is a rational number $\varepsilon, 0 < \varepsilon < \delta$ such that every real continuous function of n variables, $f(x)$, defined on I^n, can be represented as

$$f(x) = \sum_{j=1}^{2n+1} \chi (\sum_{i=1}^{n} \lambda^i \psi(x_i + \varepsilon(j-1)) + j - 1)$$

(2.81)

where the function χ is real and continuous and λ is an independent constant.

Although this theorem of Kolmogorov was powerful, in the beginning it was not found of much use in proving other important theorems. Interest in the theorem revived when one started searching for piecewise-linear model descriptions. As stated already several times, in the theory of piecewise-linear techniques, one of the major goals was to obtain good model descriptions that would cover all explicit nonlinear functions. The approximation of a nonlinear function of one variable with a PL function is relative simple and can always be described in a model like *Chua*. Therefore, one often thought that, by using the theorem of Kolmogorov, it would be possible to obtain the PL model description that would cover all explicit functions.

However, up to now this general model description has not been found, mainly because the proof of the theorem was not constructive.

Kolmogorov's theorem also became important in the field of neural networks [2.18]. Here the problem was to prove how many layers in a multi-layer perceptron network would be necessary to realize any continuous function. It turned out that with at least one hidden layer any continuous function could be realized.

2.7.2. The format of the model

In the following derivation we will demonstrate that each continuous function can be *approximately realized* by a explicit piecewise linear model description with base functions of order one and two only. To that purpose, let us first define the term *approximately realized.*

Definition 2.6   *A mapping f will be **approximately realized** by $\bar{f}$ if it will always be possible at the expense of more segments (i.e. hyperplanes in the PL modeling)*

to find an arbitrary $\varepsilon > 0$ such that $\max d(f, \bar{f}) < \varepsilon$ with d the distance in the space.

In his paper [19], Funahashi proved that for any continuous mapping f and an arbitrary $\varepsilon > 0$, there exist a k-layer ($k \geq 3$) neural network whose input-output

mapping is given by $\bar{f}$ such that $\max d(f, \bar{f}) < \varepsilon$. The neurons of at least one hidden layer must have output functions that are non-constant, bounded and monotone increasing continuous functions. The output functions of the other layers represent linear additions. Thus any continuous mapping could approximately be realized with a multi-layer neural network with at least one hidden layer ($k=3$) whose output functions are sigmoid. The fineness of the approximation and the number of necessarily hidden neurons depend on the number of layers. Two hidden layers would do better than one.

In [20] it is implicitly shown that any continuous function could be approximately close represented by the following implicit PL mapping:

$$Iy + Ax + Bu + f = 0$$
$$j = Cx + Du + g \tag{2.82}$$
$$u^T j = 0, u \geq 0, j \geq 0$$

where matrix D has the following form:

$$D = \begin{pmatrix} I & 0 \\ D_2 & I \end{pmatrix} \quad D_2 \in R^{mxn} \tag{2.83}$$

The main idea behind the proof was the principal that the neural networks, used by Funahashi, could be PL modeled in a format like (2.82-2.83) [20]. To come to the explicit model description is not difficult anymore, we simple use the modulus transformation and the notations as defined in (2.43).

Let in (2.83) $n=2$, $m=3$ and define $q_i(x)$ as $C_{\cdot i} x + g_i$ as is done in (2.41), then we have as state equation

$$
\begin{pmatrix} j_1 \\ j_2 \\ j_3 \\ j_4 \\ j_5 \end{pmatrix} = \begin{pmatrix} 1 & & & & \\ & 1 & & & \\ d_{31} & d_{32} & 1 & & \\ d_{41} & d_{42} & & 1 & \\ d_{51} & d_{52} & & & 1 \end{pmatrix} u + \begin{pmatrix} q_1(x) \\ q_2(x) \\ q_3(x) \\ q_4(x) \\ q_5(x) \end{pmatrix}
$$

It is obvious that, from the complementary of u and j, the first two rows in this system can consequently be written in terms of base functions of order one, as:

$$
u_i = \lfloor -q_i(x) \rfloor, i = 1, 2
$$

Each following row will end up in a (degenerated) base function of the second order:

$$
u_i = \lfloor -q_i(x) - d_{i1} \lfloor -q_1(x) \rfloor - d_{i2} \lfloor -q_2(x) \rfloor \rfloor, i = 3, 4, 5
$$

More general, because matrix D_2 is situated below the diagonal of the state matrix in front of u, we will always have a set of independent hyperplanes, which can be realized with base functions of order one. The rows out of D_2 are only linear combinations of these base functions and thus will end up in base functions of order two.

In section 2.5 we saw that for a complete lower triangular matrix D the base functions end up in order k with k the rank of D. However, to cover all explicit continuous functions, only the first two base functions are necessary. More hidden layers will result in a more accurate mapping of the neural network. It is easily to see that for k hidden layers, the D matrix in the state equation will look like

$$D = \begin{pmatrix} I & & & & \\ D_{21} & I & & & \\ D_{31} & D_{32} & I & & \\ & & & \ddots & \\ D_{k,1} & D_{k,2} & \cdots & D_{k,k-1} & I \end{pmatrix} \qquad (2.84)$$

and thus in higher order base functions.

It must now become clear that the base functions, as defined in section 2.5, nicely cover the theorem of Funahashi. Each extra hidden layer results in a higher order base function and in section 2.5 it was already explained that models based on those base functions would cover a broad class of explicit functions. Now it turns out that with that set of base functions, any continuous mapping can be *approximately realized.*

In principle the theorem of Kolmogorov means that any continuous function can be written as a double nesting of functions of one variable. It is known from the literature (see for instance [2.18]) that a continuous function on R can be approximated by a set of monotone increasing functions in one dimension. This means that such a function can be approximated by an explicit PL model using first order base functions and thus the function ψ from (2.80) can be realized by such first order base functions. Consider also the fact that our $\lfloor\ \rfloor$-operator is a monotone increasing function, which is exactly the property of ψ. The same holds for the function χ. However, the latter function will influence the first, ending up with a summation of base functions of order one and two in one dimension. Using (2.81) we will always have an arbitrary $\varepsilon > 0$ such that

$$\left| \sum_{j=1}^{2n+1} \chi\left(\sum_{i=1}^{n} \lambda^i \psi(x_i + \varepsilon(j-1)) + j - 1\right) - PL(x) \right| < \varepsilon$$

with *PL(x)* the piecewise linear model with only base functions of order one and two. This means that we have also a lower bound for the number of base functions. To approximate sufficiently close, at least n first order and *2n+1* second order base functions are necessary. The actual numbers will be larger due to the number of segments used to realize the output function.

2.8. Conclusions

In this chapter an overview was given of the most common model descriptions to describe piecewise linear continuous functions. The model descriptions can be implicit or explicit. So far, one of the implicit model descriptions seems to be the most powerful, it covers the largest class of functions. Therefore, this model is used to find that explicit model description to cover all explicit functions. To that purpose base functions were derived, which already gave an extension to the class of explicit model descriptions. Besides that, with the base functions, all already presented explicit model descriptions could be covered. Finally, we showed that only the first two base functions are necessary to approximately realize each explicit function. However, the better the approximation, the more base functions one needs. Because the proof of the necessity of number and order of base functions is existential, it is currently not known how to obtain those base functions for a given function description.

CHAPTER **3**

SOLUTION ALGORITHMS

Given an implicit piecewise linear model description, one needs a solution algorithm to find the corresponding output vector for a certain input vector. This problem is known as the Linear Complementary Problem and many solution techniques are available. In this chapter the most commonly used methods are discussed.

3.1. The Linear Complementary Problem

In the previous chapter several explicit and implicit piecewise linear model descriptions were treated. It was also shown that the explicit description could be assembled into the most general implicit model description, namely *Bok1,*

$$y = Ax + Bu + f$$
$$j = Cx + Du + g \qquad (3.1)$$
$$u \geq 0, j \geq 0, u^T j = 0$$

Obvious for the explicit PL model description, when an input vector is given, the corresponding output vector is immediately available by solving the absolute-sign operators. However, for the model (3.1) this does not hold. The state vectors u and j have to be find in order to solve the system equation $y=Ax+Bu+f$ for a given input vector x. For a given x the second equation in (3.1) can be redefined as

$$j = Du + q$$
$$u \geq 0, j \geq 0, u^T j = 0 \qquad (3.2)$$

with $q = Cx + g$. The problem is to find the vectors u and j consistent with (3.2) and the problem is known as the Linear Complementary Problem (LCP). The solution to this problem is the key operation in the evaluation of a PL-function based on (3.1). We refer here also to definition 2.4. The LCP is known as a basic problem for quite some time already and is mainly studied for applications in game theory and economics.

In the past 20 years a number of algorithms have been developed for solving the LCP, which in its general form is known to be an NP-complete problem. The solution can be found by going through all possible, pivotisations of the matrix D which number is exponential in the dimension of D. An in general more efficient approach is to construct algorithms which use an extension of a local solution estimate to find the required result.

Over the years several algorithms have been developed to solve the problem and they can mainly be categorized into three groups:

- *Homotopy Algorithms:* Algorithms as developed by Lemke [21], Katzenelson [22], van de Panne [23] belong to this class of pivoting algorithms. Homotopy methods add an additional parameter, say λ, to the problem and then construct a solution path in the extended solution space [24]. In general, suppose we have to solve the problem $f(x, p) = 0$ for which we already know a priori the particular solution $x_0 = x(p_0)$. This problem can now be reformulated into a one parameter problem $\bar{f}(x, p, \lambda) = f(x, p_0 + \lambda(p - p_0))$ with known solution $f(x_0, p_o) = 0$. The method starts with $\lambda = 0$ and continuously increases the value of λ until $\lambda = 1$, the situation for which the solution was to be found. The method constructed a solution $x(\lambda)$ along the path $0 \leq \lambda \leq 1$.

- *Iterative Algorithms*: These methods solve some equivalent multidimensional optimization problem. This optimization problem is most often quadratic [25]. Problem (3.2) can be reformulated as minimize $\frac{1}{2} x^T Dx + q^T x$ under the condition $x \geq 0$ which yields a solution satisfying (3.2). The required solution can be obtained by applying efficient gradient search methods from the nonlinear optimization theory.

- *Contraction Algorithms*: The algorithms in this class solve some equivalent nonlinear algebraic problem by iteration using for example contraction or Newton-Raphson iteration. One important member of this class is the modulus algorithm [2]. This method will yield a polynomial solution algorithm for matrix D from certain limited class such as e.g. positive definitive matrices.

Before paying attention to several solution algorithms, it would be of interest to investigate the conditions under which (3.1) will yield a solution for a given q and whether or not this solution will be unique. It is obvious from (3.1) that for $q \geq 0$ the solution immediately can be found by inspection as $j = q, u = 0$. In the case that some entries of q will have a negative sign, one can try to perform a sequence of diagonal pivoting steps on the linear relation (3.1) in such way that the result can be written as

$$w = \overline{D}z + \overline{q}$$
$$w \geq 0, z \geq 0, w^T z = 0, \overline{q} \geq 0$$

(3.3)

Then again $w = \overline{q}, z = 0$ will solve (3.3) and thus also yield a solution for (3.1). For any positive diagonal element of $\overline{D}$ one may easily see that a Gauss-Jordan pivoting step using such an element as a pivot, will at least change the sign of the component of the vector $\overline{q}$ in the same row, into its opposite value. In other words, for any positive diagonal element one can define pivoting steps such that all sign combinations for the entries of $\overline{q}$ can be produced, or more stronger all entries of $\overline{q}$ can be made positive. It can be proven that each sign distribution is unique which yields 2^k different combinations due to two possible signs for each of the k-vector elements. This is in agreement with the same 2^k different pivoting possibilities for the system (3.1).

This implies that a positive definitive (PD) matrix D which obviously has positive diagonal elements always leads to an existing and unique solution of the LCP (3.1) for any given vector q in a maximum of 2^k pivoting steps. This property holds even for a more broad class of matrices, called class P, of which the PD-matrices are a special case,

Definition 3.1 *A matrix D belongs to class P if and only if*
$$\forall_{u \in R^k, u \neq 0} \exists_k : u_k \cdot (Du)_k > 0$$

(see also definition 2.5).

This class-P was first defined by Fiedler and Ptak as a generalization of positive definite matrices by a number of equivalent properties of which the most simple one says that the determinant of all principal minors have to be positive [26]. The value of these determinants can be related to the signs of the diagonal elements of D. The main result concerning class-P matrices and the LCP is however that the LCP has a unique solution for any vector q if and only if $D \in P$. The algorithm to derive that solution was first given by Lemke and has an exponential computational complexity. It is known that the LCP with $D \in P$ is NP-hard.

A survey of classes of matrices, suitable for the classification of the LCP, according to the paper of Karamardian is given below [27]:

- Positive definitive matrices (PD): D is of class PD if and only if
$$\forall_{u \in R^k, u \neq 0} : u^T Du > 0$$

- Positive semi-definite matrices (PSD): D is of class PSD if and only if
$$\forall_{u \in R^k} : u^T Du \geq 0$$

- P matrices: see definition 3.1
- P0 matrices (P0): D is of class P0 if and only if $\forall_{u\in R^k, u\neq 0} \exists_k : u_k \cdot (Du)_k \geq 0$
- Strict copositive matrices (SCP): D is of class SCP if and only if $\forall_{u\in R^k_+, u\neq 0} : u^T Du > 0$
- Copositive Plus matrices (CPP): D is of class CPP if and only if $\forall_{u\in R^k_+} : u^T Du > 0$ and also $\forall_{u\in R^k_+, u\neq 0, u^t Du=0} : (D+D^T)u = 0$
- Strict semi monotone matrices (SSM): D is of class SSM if and only if $\forall_{u\in R^k, u\neq 0} \exists_k : u_k \cdot (Du)_k > 0$
- L matrices (L): D is of class L if and only if $\forall_{u\in R^k, u\neq 0} \exists_k : u_k > 0 \wedge (Du)_k \geq 0$ and

  $\exists_{\Lambda, \Omega \geq 0} \forall_{u\in R^k_+, u\neq 0, Du\geq 0, u^T Du=0} : \Omega u \neq 0 \wedge (\Lambda D + D^T \Omega)u = 0$ in which Λ and Ω denote diagonal matrices

The relations between these classes can be depicted as in Fig. 3.1. A downward link means "is a superset from".

Besides the known properties related to the LCP of class-P matrices, it can be mentioned that there exist an odd number of solutions if D belongs to SSM. It can further be mentioned that for all given classes the matrices do have positive diagonal elements.

In the next several sections we will focus on solution algorithms to solve the LCP problem as defined in (3.1).

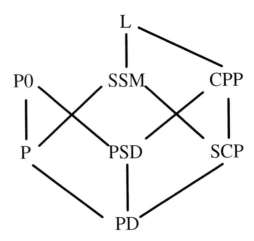

Figure 3.1. The relation between several classes of matrices

3.2. The Katzenelson algorithm

As explained in the introduction, the Katzenelson algorithm is a homotopy method. Katzenelson introduced this methods in 1965 and the method is still extensively used in piecewise linear simulation programs [22]. We will come back to this simulation aspect in chapter 4. Being a homotopy, a continuous path through the space is created by extending the LCP (3.1) according to

$$j = Du + q_0 + \lambda\left(q^* - q_0\right) \qquad (3.4)$$

where we assume that q_0 is known with $u = 0, j > 0$. We are looking for the solution for some q^*.

The homotopy parameter λ is to be increased from zero up to one. The procedure is to gradually increase parameter λ until some component j_m becomes zero, because $q_0 + \lambda\left(q^* - q_0\right) = q_m$ becomes zero. Just a small increase of λ is enough to let $u_m > 0$ to prevent j_m to become negative, which is not allowed according to the complementary conditions. Now we have to perform a pivoting operation with as result a system of equations according to (3.1). The pivot is the diagonal element D_{mm} that we assume to be positive. As a result, the variables j_m and u_m will change place and (3.1) turns into a new form given by

$$v = \overline{D}w + \overline{q}_0 + \lambda\left(\overline{q}^* - \overline{q}_0\right), \qquad v, w \geq 0, v^T w = 0 \qquad (3.5)$$

for which $w = 0$ and $v = \overline{q}_0 + \lambda_m\left(\overline{q}^* - \overline{q}_0\right)$ now will be a solution. This process of increasing λ is repeated until we reach $\lambda = 1$ in which case the solution for the LCP has been obtained.

It can be shown that λ can always be increased when the diagonal elements of D that we need as pivot are always positive. Moreover, if the matrix D belongs to class P, the Katzenelson algorithm will always find the unique solution [28].

As example consider the following problem

$$\begin{pmatrix} j_1 \\ j_2 \end{pmatrix} = \begin{pmatrix} 2 & -1 \\ 4 & 1 \end{pmatrix}\begin{pmatrix} u_1 \\ u_2 \end{pmatrix} + \begin{pmatrix} -2 \\ 1 \end{pmatrix}$$

It is obvious that $u=0$ will violate the complementary conditions, so pivoting operations have to be performed. Let us define the vector $q_0 = \begin{pmatrix} 1 & 1 \end{pmatrix}^T$ which certainly gives the solution $u=0$. Equation (3.4) can now be rewritten as

$$\begin{pmatrix} j_1 \\ j_2 \end{pmatrix} = \begin{pmatrix} 2 & -1 \\ 4 & 1 \end{pmatrix} \begin{pmatrix} u_1 \\ u_2 \end{pmatrix} + \begin{pmatrix} 1 \\ 1 \end{pmatrix} + \lambda \left\{ \begin{pmatrix} -2 \\ 1 \end{pmatrix} - \begin{pmatrix} 1 \\ 1 \end{pmatrix} \right\}$$

or

$$\begin{pmatrix} j_1 \\ j_2 \end{pmatrix} = \begin{pmatrix} 2 & -1 \\ 4 & 1 \end{pmatrix} \begin{pmatrix} u_1 \\ u_2 \end{pmatrix} + \begin{pmatrix} 1 \\ 1 \end{pmatrix} + \lambda \begin{pmatrix} -3 \\ 0 \end{pmatrix}$$

Let us increase λ until one of the components of $q_0 + \lambda(q^* - q_0) = q_m$ becomes zero, which happens at $\lambda = 1/3$ for the first entry. Performing a Gauss-Jordan pivot operation with as pivot $D_{11} = 2$ results in

$$\begin{pmatrix} u_1 \\ j_2 \end{pmatrix} = \begin{pmatrix} 1/2 & 1/2 \\ 2 & 1 \end{pmatrix} \begin{pmatrix} j_1 \\ u_2 \end{pmatrix} + \begin{pmatrix} -1/2 \\ -1 \end{pmatrix} + \lambda \begin{pmatrix} 3/2 \\ 6 \end{pmatrix}$$

Now it becomes possible to increase λ further to one without having q_m become negative again. At this point we obtain

$$\begin{pmatrix} u_1 \\ j_2 \end{pmatrix} = \begin{pmatrix} 1/2 & 1/2 \\ 2 & 1 \end{pmatrix} \begin{pmatrix} j_1 \\ u_2 \end{pmatrix} + \begin{pmatrix} 1 \\ 5 \end{pmatrix}$$

which yields the solution $(u_1 \quad j_2) = (1 \quad 5), (j_1 \quad u_2) = (0 \quad 0)$. This is indeed the solution for the given problem.

In chapter 2 we discussed piecewise linear model descriptions. It was explained that for each subspace (polytope) a linear mapping was defined. Katzenelson's algorithm, by increasing the homotopy parameter, creates a path through the subspaces. Each time when this path hits a boundary of a certain polytope, the corresponding state equation (a hyperplane) will change sign (see section 2.3) and we enter the opposite half space of the hyperplane. This means that the corresponding u and j have to be exchanged, i.e. we have to perform an update. Normally, the Katzenelson path extends through space such that it only hits one boundary at the time. The choice of the pivot is then straightforward. However, it can be possible that the path hits a corner of several polytopes.

Consider the following example

$$\begin{pmatrix} j_1 \\ j_2 \\ j_3 \end{pmatrix} = \begin{pmatrix} 1 & 1/2 & 1 \\ 1/2 & 1 & 1 \\ 0 & 0 & 1 \end{pmatrix} u + \begin{pmatrix} 1 \\ 1 \\ 2 \end{pmatrix} + \lambda \left\{ \begin{pmatrix} -1 \\ -1 \\ -2 \end{pmatrix} - \begin{pmatrix} 1 \\ 1 \\ 2 \end{pmatrix} \right\}$$

$$\begin{pmatrix} j_1 \\ j_2 \\ j_3 \end{pmatrix} = \begin{pmatrix} 1 & 1/2 & 1 \\ 1/2 & 1 & 1 \\ 0 & 0 & 1 \end{pmatrix} u + \begin{pmatrix} 1 \\ 1 \\ 2 \end{pmatrix} + \lambda \begin{pmatrix} -2 \\ -2 \\ -4 \end{pmatrix}$$

where we assumed to find a solution for $q = \begin{pmatrix} -1 & -1 & -2 \end{pmatrix}^T$. For $\lambda = 1/2$ the pivot is obtained in all the three equation. We have hit the corner or intersection point of the three hyperplanes that divide the space into 6 regions. After updating the first equation, we have

$$\begin{pmatrix} u_1 \\ j_2 \\ j_3 \end{pmatrix} = \begin{pmatrix} 1 & -1/2 & -1 \\ 1/2 & 3/4 & 1/2 \\ 0 & 0 & 1 \end{pmatrix} \begin{pmatrix} j_1 \\ u_2 \\ u_3 \end{pmatrix} + \begin{pmatrix} -1 \\ 1/2 \\ 2 \end{pmatrix} + \lambda \begin{pmatrix} 2 \\ -1 \\ -4 \end{pmatrix}$$

from which we can see that $\lambda = 1/2$ is still the solution, but now only for the second and third equation. We have passed the first hyperplane. Performing the other updates results in

$$\begin{pmatrix} u_1 \\ u_2 \\ u_3 \end{pmatrix} = \begin{pmatrix} 4/3 & -2/3 & -2/3 \\ -2/3 & 4/3 & -2/3 \\ 0 & 0 & 1 \end{pmatrix} \begin{pmatrix} j_1 \\ j_2 \\ j_3 \end{pmatrix} + \begin{pmatrix} 2/3 \\ 2/3 \\ -2 \end{pmatrix} + \lambda \begin{pmatrix} -4/3 \\ -4/3 \\ 4 \end{pmatrix}$$

after which we have to conclude that again we have to update over the first two equations. We might ask ourselves what happened? We assumed that we had to pass all three hyperplanes in the corner point, but this was wrong. We were at the correct site for the first two planes and we only had to pass the third one. If the start situation was updated directly using the pivot from the third equation we had obtained

$$\begin{pmatrix} j_1 \\ j_2 \\ u_3 \end{pmatrix} = \begin{pmatrix} 1 & 1/2 & -1 \\ 1/2 & 1 & 1 \\ 0 & 0 & 1 \end{pmatrix} \begin{pmatrix} u_1 \\ u_2 \\ j_3 \end{pmatrix} + \begin{pmatrix} -1 \\ -1 \\ -2 \end{pmatrix} + \lambda \begin{pmatrix} 2 \\ 2 \\ 4 \end{pmatrix}$$

This result was also obtained by performing again two updates over the last result. Obviously we could not foresee which updates we had to perform, which is a problem when hitting a corner. Instead of performing 5 updates only one was sufficient. The only way to avoid this problem is to modify the path by a small perturbation. Choosing $q = (3/2 \quad 1/2 \quad 2)^T$ as starting point, the first and only update is found at $\lambda = 1/2$ in the third equation, after which the answer can be found

$$
\begin{pmatrix} j_1 \\ j_2 \\ u_3 \end{pmatrix} = \begin{pmatrix} 1 \\ 1 \\ 2 \end{pmatrix}
$$

This is the same answer as obtained by performing the five updates. The conclusion is the following. When a corner point is reached, noticed due to several pivots at the same time are possible, it is better to perturb the initial condition or change the direction of the path at that point, rather then to proceed. The corner problem was for the first discussed in the work of Chien and Kuh [33]. In this work they proved that it is possible to choose a path from a given starting point such that this line will never reach or cross a corner. From the computational point of view it is certainly impractical to compute that path. And since it is not a frequent event that the solution curve will hit a corner, the suggested option to perturb slightly the path will do.

The algorithm of Katzenelson is still widely used due to is simplicity of implementation. We refer to chapter 4 where a piecewise linear simulator will be discussed, which uses this algorithm with some modifications. We will now conclude with a simple electrical network example.

As an example consider a fairly simple network, consisting of a linear resistor in series with a nonlinear resistor that has a characteristic as defined in Fig. 3.2 and for which the model is given by

$$
i + (-1)v + \left(\frac{3}{2} \quad -\frac{3}{2} \right) u = 0
$$

$$
j = \begin{pmatrix} -1 \\ -1 \end{pmatrix} v + Iu + \begin{pmatrix} 1 \\ 2 \end{pmatrix}
$$

$$
u, j \geq 0 \qquad u^T j = 0
$$

This network is excited by a voltage source E. The topological relation yields

$$
E = Ri + v
$$

For this network we intend to find the DC-operating point for $E = 9$ V and $R = 4$ Ohm. According to the theory in section 1 we can write the complete network in

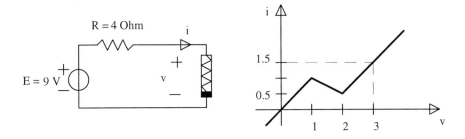

Figure 3.2. Network with nonlinear resistor with piecewise linear behavior

terms of its input variable E and its output variable i by combining the topological equations with the description of the piecewise linear resistor, yielding

$$i + \left(-\frac{1}{5}\right)E + \left(\frac{3}{10} \quad -\frac{3}{10}\right)u = 0$$

$$j = \binom{-1}{-1}E + \binom{-4}{-4}i + Iu + \binom{1}{2}$$

where we leave out the complementary conditions for convenience. Because of the definition of the elements of the network, $(i, E_0) = (0,0)$ is a solution of the network. However, we intend to obtain the DC-operating point for $E_e = 9$ and therefore we may define the homotopy path as $E = E_o + \lambda(E_e - E_0) = \lambda 9$. We are now able to rewrite the state equation into a form similar to (3.4) yielding

$$j = \binom{-\frac{1}{5}}{-\frac{1}{5}}\lambda 9 + Iu + \binom{1}{2}$$

Note that the LCP matrix is the identity matrix and thus of class P. Katzenelson's algorithm will always obtain a solution. Increasing λ to let $u_m > 0$ to prevent j_m to become negative results in $\lambda = \frac{5}{9}$ for the first state equation. Let u_1 and j_1 inter change and the update will leads us to

$$i + \left(-\frac{1}{2}\right)E + \left(-\frac{3}{2} \quad \frac{3}{2}\right)u + \frac{3}{2} = 0$$

$$j = \binom{1}{-1}E + \binom{-4}{4}i + Iu + \binom{-1}{2}$$

where u_1 and j_1 inter changed names. Note that $\lambda = \frac{5}{9}$ means that $E = 5$, $i = 1$ and therefore $v = 1$ which is indeed a breakpoint of the nonlinear resistor characteristic (see Fig. 3.2). By further increasing E the diode in the second branch of the sub network representing the nonlinear resistor starts to conduct as v increases. The complete network topology will now change and is described by the new mapping equation. For this new situation we obtain

$$ j = \begin{pmatrix} -1 \\ 1 \end{pmatrix} \lambda 9 + Iu + \begin{pmatrix} 5 \\ -4 \end{pmatrix} $$

from which it can be observed that we may not increase the homotopy parameter λ further. The alternative is to decrease this parameter and hoping that we may increase it afterwards to finally reach $\lambda = 1$. It can be proved that this extension to the original method of Katzenelson is allowed [33]. Doing so, we obtain $\lambda = \frac{4}{9}$ in the second state equation which corresponds to the diode in third branch of the sub network representing the nonlinear resistor starting to conduct. Pivoting and updating the model results in

$$ i + \left(-\frac{1}{5} \right) E + \left(\frac{3}{10} \quad -\frac{3}{10} \right) u + \frac{3}{10} = 0 $$

$$ j = \begin{pmatrix} 1 \\ 1 \end{pmatrix} E + \begin{pmatrix} -4 \\ -4 \end{pmatrix} i + Iu + \begin{pmatrix} -1 \\ -2 \end{pmatrix} $$

and the algorithm yields

$$ j = \begin{pmatrix} \frac{1}{5} \\ \frac{1}{5} \end{pmatrix} \lambda 9 + Iu + \begin{pmatrix} \frac{1}{5} \\ -\frac{4}{5} \end{pmatrix} $$

from which it is clear that we may increase the homotopy parameter reaching $\lambda = 1$. We now have obtained the DC-operating point of this network, $(i, E) = \left(\frac{3}{2}, 9 \right)$ and the voltage over the nonlinear resistor is $v = 3 \ V$.

3.3.   The modulus algorithm

One of the most elegant solution methods for solving the LCP relies on a transformation of the co-ordinate system. This method is called the modulus algorithm [2] and is a contraction method. Consider Fig. 3.3, where the condition on u and j for a single pair u, j are drawn. If a rotation of the co-ordinate system over

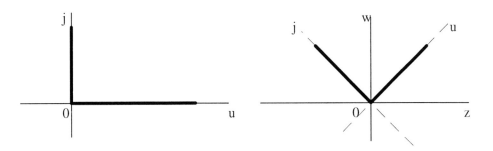

Figure 3.3. Rotation of the system u,j.

$45°$ is performed by changing from the pair u,j to the pair z,w according to the affine mapping

$$w = u + j$$
$$z = u - j$$
(3.6)

it appears that the conditions on u,j now take the shape of a modulus or absolute-sign function in the w,z co-ordinate system, i.e. the complementary conditions on u,j are condensed into the single relation

$$w = |z|$$
(3.7)

Hence the complementary pair u,j is transformed by (3.6) into (3.7). Substitution of (3.7) into (3.6) and solving u and j yields

$$u = (|z| + z) / 2$$
$$j = (|z| - z) / 2$$
(3.8)

that is known to be the modulus transformation. This transformation was already introduced in chapter 2, definition 2.3, and was used to show the relations between several model descriptions.

The in (3.8) defined '*rotation*' transforms a complementary vector pair u,j into a single vector z and the other way around. This transformation is a one-to-one mapping. Substitution of (3.8) into (3.2) yields

$$|z| - z = D(|z| + z) + 2q$$
(3.9)

that after reordering results into the set of nonlinear equations

$$z = (I + D)^{-1}(I - D)|z| - 2(I + D)^{-1}q$$

or in the general format

$$z = M|z| + b \qquad\qquad (3.10)$$

The *modulus transformation* has changed the LCP (3.2) into the nonlinear vector equation (3.10). The solution to such an equation is most easily derived by a so-called *successive substitution method*. This method sets up the following iteration for (3.10):

$$z_{k+1} = M|z_k| + b, \qquad \text{with } z_0 = 0, k = 0,1,2,\dots \qquad\qquad (3.11)$$

If $\|M\| < 1$, where we use the absolute norm and $\|A\| = \sup_{x \neq 0} \dfrac{\|Ax\|}{\|x\|}$, this iterative

procedure will certainly converge to a solution for $k \to \infty$. Since $\|M\|$ does not depend on the iterative sequence of z it follows that $f(z) = M|z| + b$ is Lipschitz bounded. Thus

$$\|f(z_1) - f(z_2)\| \le \|M\| \cdot \|z_1 - z_2\| \qquad\qquad (3.12)$$

and then if $\|M\| < 1$, the iteration $z_{k+1} = f(z_k)$ converges by the Banach contraction mapping theorem and $z^* = \lim_{k \to \infty} z_k$ is the unique solution of $z = f(z)$ [29].

Consider again the example

$$\begin{pmatrix} j_1 \\ j_2 \end{pmatrix} = \begin{pmatrix} 2 & -1 \\ 4 & 1 \end{pmatrix} \begin{pmatrix} u_1 \\ u_2 \end{pmatrix} + \begin{pmatrix} -2 \\ 1 \end{pmatrix}$$

that using the modulus transformation can be written as

$$z = 1/5 \begin{pmatrix} -3 & 1 \\ -4 & -2 \end{pmatrix} |z| + \begin{pmatrix} 3/5 \\ -11/5 \end{pmatrix}$$

Starting with $z_0 = (0,0)^T$ the first two iterations yield

$$z_1 = \begin{pmatrix} 3/5 \\ -11/5 \end{pmatrix}, \quad z_2 = \begin{pmatrix} 0.68 \\ -3.56 \end{pmatrix}$$

and after 26 iterations the method converge to

$$z_{26} = z^* = \begin{pmatrix} 1 \\ -5 \end{pmatrix}$$

for which we obtain using (3.8) $(u_1 \quad j_2) = (1 \quad 5), (j_1 \quad u_2) = (0 \quad 0)$, the same result as using the Katzenelson algorithm.

The complexity of the above mentioned algorithm is exponential. However, when D is of class PD *and* the matrix is symmetric, it can be proven that the algorithm is of polynomial complexity [2]. Let $D \in R^{nxn}$ and k the condition number of D. Then the number N of multiplications or additions is asymptotically bounded by

$$N \sim \frac{1}{7} K^{1/2} n^{7/2} \tag{3.13}$$

The proof is based on the fact that a symmetric PD matrix has positive real eigen values and therefore M in (3.10) has also real eigen values μ_i that also satisfy $|\mu_i| < 1$.

3.4.   Lemke and van de Panne algorithms

The most familiar algorithm to solve the LCP problem is the well-known Lemke algorithm [21]. Lemke proposed his method in 1965 but presented a powerful extension to his algorithm in 1968. Consider again the basic problem (3.2). The principal of the method is to extend the vector q by a multiple λ of some positive vector e yielding

$$\begin{pmatrix} j_\lambda \\ \cdots \\ j \end{pmatrix} = \begin{pmatrix} 0 & \vdots & 0 \\ \cdots & \vdots & \cdots \\ e & \vdots & D \end{pmatrix} \begin{pmatrix} \lambda \\ \cdots \\ u \end{pmatrix} + \begin{pmatrix} 0 \\ \cdots \\ q \end{pmatrix} \rightarrow w = a + Mz, w \geq 0, z \geq 0, w^T z = 0 \tag{3.14}$$

Obviously, if after some pivoting steps a solution of (3.14) is obtained for which $\lambda = 0$, then the components w and z of that solution also satisfy (3.2), i.e. the original problem. Let us now consider a value λ such that

$$\lambda = \min_k \left(-\frac{a_k}{e_k} \middle| a_k < 0 \right) \tag{3.15}$$

then for this particular row p of our system, $w_p=0$. Next, system (3.14) is pivoted on element $M_{p,1}$. Due to this pivoting, λ is moved from the z-vector to the w-vector, but on row $r \neq p$. Therefore, we have *almost* complementary conditions. To correct the system, we have to perform an update over $\overline{M}_{rp}$ where the bar means the update of matrix M over $M_{p,1}$. We have to repeat these two pivoting steps until λ can be chosen zero. This means we satisfy the complementary conditions or no component of w can be chosen. In the latter case no solution is obtained. Lemke showed that this can not arise for $D \in P$ whereas Dantzig and Cottle proved this for class SSM [30] and finally Eaves extended these results to class L matrices [31]. The convergence proof for the algorithm relies on the fact that all visited states are complementary. Moreover, all states have a unique successor as well as a unique predecessor, except the starting state and final state which have only a successor and predecessor respectively.

Consider again the example

$$\begin{pmatrix} j_1 \\ j_2 \end{pmatrix} = \begin{pmatrix} 2 & -1 \\ 4 & 1 \end{pmatrix} \begin{pmatrix} u_1 \\ u_2 \end{pmatrix} + \begin{pmatrix} -2 \\ 1 \end{pmatrix}$$

that can be extended with the vector $e = (1,1)^T$ resulting in

$$\begin{pmatrix} j_1 \\ j_2 \end{pmatrix} = \begin{pmatrix} 2 & -1 \\ 4 & 1 \end{pmatrix} \begin{pmatrix} u_1 \\ u_2 \end{pmatrix} + \begin{pmatrix} -2 \\ 1 \end{pmatrix} + \lambda \begin{pmatrix} 1 \\ 1 \end{pmatrix}$$

or

$$\begin{pmatrix} j_\lambda \\ j_1 \\ j_2 \end{pmatrix} = \begin{pmatrix} 0 & 0 & 0 \\ 1 & 2 & -1 \\ 1 & 4 & 1 \end{pmatrix} \begin{pmatrix} \lambda \\ u_1 \\ u_2 \end{pmatrix} + \begin{pmatrix} 0 \\ -2 \\ 1 \end{pmatrix}$$

This is the starting point for the Lemke algorithm. According to (3.15) $\lambda = 2, p = 2$. Now choosing the pivot $M_{2,1} = 1$ yields

$$\begin{pmatrix} j_\lambda \\ \lambda \\ j_2 \end{pmatrix} = \begin{pmatrix} 0 & 0 & 0 \\ 1 & -2 & 1 \\ 1 & 2 & 3 \end{pmatrix} \begin{pmatrix} j_1 \\ u_1 \\ u_2 \end{pmatrix} + \begin{pmatrix} 0 \\ 2 \\ 3 \end{pmatrix}$$

which is an almost complementary condition when $z=0$. Pivoting on $\overline{M}_{2,2} = -2$ in order to exchange λ results in

$$\begin{pmatrix} j_\lambda \\ u_1 \\ j_2 \end{pmatrix} = \begin{pmatrix} 0 & 0 & 0 \\ 1/2 & -1/2 & 1/2 \\ 2 & -1 & 1 \end{pmatrix} \begin{pmatrix} j_1 \\ \lambda \\ u_2 \end{pmatrix} + \begin{pmatrix} 0 \\ 1 \\ 5 \end{pmatrix}$$

$$\begin{pmatrix} j_\lambda \\ u_1 \\ j_2 \end{pmatrix} = \begin{pmatrix} 0 & 0 & 0 \\ -1/2 & 1/2 & 1/2 \\ -1 & 2 & 1 \end{pmatrix} \begin{pmatrix} \lambda \\ j_1 \\ u_2 \end{pmatrix} + \begin{pmatrix} 0 \\ 1 \\ 5 \end{pmatrix}$$

which can satisfy the complementary conditions for $\lambda = 0$ and gives us the already known solution $(u_1 \quad j_2) = (1 \quad 5), (j_1 \quad u_2) = (0 \quad 0)$.

From the example it can be seen that also off-diagonal pivots are used in contrast to the algorithm of Katzenelson. This makes the implementation of this algorithm more difficult, especially when sparse matrix techniques are to be used. This aspect occurs when simulating large circuits, as will be discussed in chapter 4.

In 1974, van de Panne published an algorithm to overcome two problems in the Lemke algorithm [23]. First, only diagonal elements are used as pivots and second a powerful feasibility check can be done during the run of the algorithm. The price to be paid is a more complex algorithm. For the remaining the two methods are equivalent, which also means that they can handle the same class of matrices. For reasons of complexity the treatment of this algorithm will be skipped. But related documentation can be found in [13,32].

3.5. Other solution algorithms

There are many more solution algorithms to solve the linear complementary problem. Consider for instance the class of iterative algorithms. The LCP of (3.2) is equivalent to finding vectors $u, j \in R_+^n$ such that for some $\alpha > 0$:

$$f(u, j) = \alpha j \cdot u + \frac{1}{2} (Du + q - j) \cdot (Du + q - j) = 0 \tag{3.16}$$

which is, if feasible, a minimum of f. The equivalence is obvious: $f(u,j)=0$ for all pairs u,j that solve (3.2) and otherwise $f(u,j)>0$. Define first $x = (u \quad j)^T \in R_+^{2n}$ with n the dimension of the state vectors. Define the gradient $g(x)$ and the Hessian H of $f(u, j) = f(x)$ by:

$$g(x) = g(u,j) = \alpha \begin{pmatrix} j \\ u \end{pmatrix} + \begin{pmatrix} D^T(Du+q-j) \\ -(Du+q-j) \end{pmatrix}$$

$$H = (h_{km}) = \alpha \begin{pmatrix} 0 & I \\ I & 0 \end{pmatrix} + \begin{pmatrix} D^T D & -D^T \\ -D & I \end{pmatrix}$$

(3.17)

Cryer [25] proposed the following solution strategy to minimize $f(x)$. Define the iteration sequence $r = 0,1,2,\ldots$ and for each iteration point k the vectors $x_r \in R_+^{2n}$. Further for each k we define for $m = 0,1,2,\ldots,2n$ the vectors $x^{r,m} \in R_+^{2n}$ such that for $x_0 \in R_+^{2n}$ as a given starting point of the algorithm, we develop the sequence

$$x_s^{r,m} = \begin{cases} x_s^r & \text{for } 1 \le s \le m \\ x_s^{r-1} & \text{for } m+1 \le s \le 2n \end{cases}$$

(3.18)

with

$$x_s^{r,m} = \max\left(0, x_s^{r-1} - \omega g_s\left(x^{r,s-1}\right)/h_{ss}\right)$$

in which ω an arbitrary relaxation parameter with $0 < \omega < 2$. It can be proven that the sequence x^r will converge for D of class SSM and that limit point is the solution of the LCP if D of class P [32].

An other class of methods is based on integer labeling [34-36]. This technique is mainly used to solve equations of the form $x = f(x)$ and appear to be very efficient if the matrices involved are sparse. The algorithms are based on two steps: triangulation and labeling. The space of the state variables u is divided into a fixed grid, with each interval defined by an integer 'label'. These labels are then used to indicate some property of u with respect to the LCP to be solved. If each vertice of a certain subspace, called a simplex, carries a different label, then this simplex contains the solution of the problem.

An other interesting approach to solve the LCP was proposed by de Moor [37] and independently by Leenaerts [39]. The basic idea is to rewrite the LCP is a set of linear equalities that has to be solved,

$$d_{11}u_1 + d_{12}u_2 + \ldots + d_{1n}u_n - j_1 + q_1 s = 0$$

$$\vdots$$

$$d_{n1}u_1 + d_{n2}u_2 + \ldots + d_{nn}u_n - j_n + q_n s = 0$$

(3.19)

$$u_i \ge 0, j_i \ge 0, i = 1,2,\ldots,n$$

$$s = 1$$

with s a slack variable. The solutions of this system can be obtained by a technique proposed by Tschernikow [39]. A detailed outline of this technique can also be found in [40]. To be consistent with the complementary conditions $u^T j = 0$ the method is slightly modified. We will demonstrate the working of this technique to our example

$$\begin{pmatrix} j_1 \\ j_2 \end{pmatrix} = \begin{pmatrix} 2 & -1 \\ 4 & 1 \end{pmatrix} \begin{pmatrix} u_1 \\ u_2 \end{pmatrix} + \begin{pmatrix} -2 \\ 1 \end{pmatrix}$$

which according to (3.19) can be rewritten as

$$\begin{pmatrix} 2 & -1 & -1 & 0 & -2 \\ 4 & 1 & 0 & -1 & 1 \end{pmatrix} \begin{pmatrix} u_1 \\ u_2 \\ j_1 \\ j_2 \\ s \end{pmatrix} = 0 \rightarrow Ax = 0, x \geq 0$$

The method is based on a sequence of matrix manipulations, starting with the initial tableau,

$$\begin{pmatrix} 1 & 0 & 0 & 0 & 0 & | & 2 & 4 \\ 0 & 1 & 0 & 0 & 0 & | & -1 & 1 \\ 0 & 0 & 1 & 0 & 0 & | & -1 & 0 \\ 0 & 0 & 0 & 1 & 0 & | & 0 & -1 \\ 0 & 0 & 0 & 0 & 1 & | & -2 & 1 \end{pmatrix}$$

The left part of the matrix is the unity matrix, while the columns of the right part are the rows of our initial problem. Tschernikow's method consists of two steps performed in a sequence. The first step is that all rows with a zero entry in the first column of the right part of the matrix are transferred to the new tableau matrix. In this situation it applies for row 4. Secondly, only combinations of rows can be transferred to the new matrix for which the linear positive combination leads to a zero entry in the first column of the right part. This holds for all combinations, leading to

$$\begin{pmatrix} 0 & 0 & 0 & 1 & 0 & | & 0 & -1 \\ 1 & 2 & 0 & 0 & 0 & | & 0 & 6 \\ 1 & 0 & 2 & 0 & 0 & | & 0 & 4 \\ 1 & 0 & 0 & 0 & 1 & | & 0 & 5 \end{pmatrix}$$

But to be consistent with the complementary conditions, row 3 must be removed, it violates $u_1 j_1 = 0$ which can be seen in the left part of the matrix where now each column represents one variable of vector x. Applying the same technique on the resulting matrix yields

$$\begin{pmatrix} 1 & 2 & 0 & 5 & 0 & | & 0 & 0 \\ 1 & 0 & 0 & 5 & 1 & | & 0 & 0 \end{pmatrix}$$

Now the algorithm terminates because all columns in the right part have zero entries. Because $s=1$ must hold only the second row yields, and indeed represents the solution. It can be proven that this always holds if the LCP matrix D is of class P. The disadvantage of this method is its complexity, the advantage is that even multiple solutions can be obtained. This advantage will be used to find multiple solutions of circuits as will be treated in chapter 6.

3.6.   Explicit solutions

Among the methods mentioned previously, many of them allow for an explicit description of PL-functions in terms of modulus functions. They all appear to rely on the fact that the corresponding state model description includes an LCP-problem with a lower triangular matrix, which can explicitly be solved. Reconsider chapter 2, where we showed that almost all model descriptions have a lower triangular state matrix, when reformatted into the general model *Bokh1*. This property means that it should be possible to solve the related LCP problem explicit. In section 2.6.1 we already showed that even a full matrix of class-P has the property that it can be reformatted into a lower triangular matrix, meaning the LCP can explicitly be solved. In section 2.6.1 we said that via an induction step one could prove that the mentioned property for class-P matrices hold. Here we will prove this assumption step.

In this section we will present a construction for solutions in explicit form for any class-P LCP. Let us start with the one-dimensional situation, i.e. $P \in R$ and the LCP of (3.2) looks like

$$j = du + q$$
$$u, j \geq 0, uj = 0$$

(3.20)

with d and q scalars. Because $d \in P$, we know that $d>0$ and hence we easily observe

$$
\begin{aligned}
q > 0 &\Rightarrow j = q, u = 0 \\
q \leq 0 &\Rightarrow j = 0, u = -q/d
\end{aligned}
\tag{3.21}
$$

Equation (3.20) can be written in a single explicit formula

$$
j = \lfloor q \rfloor \text{ and } u = \lfloor -q/d \rfloor
\tag{3.22}
$$

which is a basic step in the construction of the solution.

Let us rewrite the n-dimensional LCP into the format

$$
\begin{aligned}
j_1 &= D_{11} u_1 + \hat{d} \\
\hat{j} &= D_{\bullet 1} u_1 + \overline{D}\hat{u} + \overline{q}
\end{aligned}
\tag{3.23}
$$

with

$$
\begin{aligned}
\hat{j} &= \begin{pmatrix} j_2 & \cdots & j_n \end{pmatrix}^T, \hat{u} = \begin{pmatrix} u_2 & \cdots & u_n \end{pmatrix}^T \\
d &= \sum_{i=2}^{n} D_{1i} u_i + q_1 = d(u) \\
\overline{D} &= \begin{pmatrix} D_{22} & \cdots & D_{2n} \\ \vdots & \ddots & \vdots \\ D_{n2} & \cdots & D_{nn} \end{pmatrix} \\
\overline{q} &= \begin{pmatrix} q_2 & \cdots & q_n \end{pmatrix}
\end{aligned}
\tag{3.24}
$$

and define $\hat{d} = d(\hat{u})$. The unique solution of (3.23) will be partitioned and denoted as $j = \begin{pmatrix} j_1 \\ \hat{j} \end{pmatrix}$ and $u = \begin{pmatrix} u_1 \\ \hat{u} \end{pmatrix}$. Let us also define an other $(n-1)$-dimensional LCP, to be derived from (3.23) by taking $u_1 = 0$

$$
\tilde{j} = \overline{D}\tilde{u} + \overline{q}
\tag{3.25}
$$

also having unique solutions, because also $\overline{D} \in P$. In a similar way as $\hat{d}$ we define $\tilde{d} = d(\tilde{u})$.

Starting from the unique solution u and j of (3.23), determine the value of $\hat{d}$. Suppose $\hat{d} \geq 0$ then from (3.23) $j_1 = \hat{d}, u_1 = 0$. This result implies, comparing (3.23) with (3.25), that $\hat{j} = \tilde{j}, \hat{u} = \tilde{u}$. Consequently, this equivalence yields that $\hat{d} = d(\hat{u}) = d(\tilde{u}) = \tilde{d} \geq 0$ in which $\tilde{d}$ can be derived solely from the solutions of the reduced LCP (3.25). Using (3.21-3.22) we now have

$$j_1 = \hat{d} = \tilde{d} = \lfloor \hat{d} \rfloor = \lfloor \tilde{d} \rfloor \tag{3.26}$$

Along the same lines, supposing $\tilde{d} \geq 0$ again leads to (3.26) and hence we have

$$\left(\hat{d} \geq 0 \text{ or } \tilde{d} \geq 0 \right) \Rightarrow \hat{d} = \tilde{d} \geq 0 \tag{3.27}$$

Now suppose that $\tilde{d} < 0$ then the following property holds:

$$\tilde{d} < 0 \Rightarrow \hat{d} < 0 \tag{3.28}$$

This remarkable property can easily be proved. Suppose that $\tilde{d} < 0$ and $\hat{d} \geq 0$. Then from $\hat{d} \geq 0$ and (3.27) we have $\tilde{d} = \hat{d}$ and consequently $\tilde{d} \geq 0$, which is a contradiction with $\tilde{d} < 0$. Hence, property (3.28) holds. Therefore when $\tilde{d} < 0$ we have the remarkable property $sign(\hat{d}) = sign(\tilde{d})$. In this case, again $j_1 = \lfloor \hat{d} \rfloor = \lfloor \tilde{d} \rfloor = 0$ holds.

Combining (3.27) with (3.28) finally yields, independently of the sign of $\tilde{d}$

$$j_1 = \lfloor \tilde{d} \rfloor \tag{3.29}$$

Hence with (3.29) the solution of the n-dimensional LCP (3.23) is expressed in terms of the solutions of the reduced order $(n-1)$-dimensional problem (3.25). This process can be repeated until a scalar problem is obtained that is solved by (3.22). In this way the solution can explicitly be written down in a recursive fashion.

Observe that intermediate values of u_k and j_k obtained in the flow of the above procedure are only valid within the formula for the final solution.

Note the remarkable property that always $\lfloor \hat{d} \rfloor = \lfloor \tilde{d} \rfloor$, i.e. the value of $\lfloor \hat{d} \rfloor$ is fully determined by the value of $\lfloor \tilde{d} \rfloor$ of the $(n-1)$-dimensional sub problem.

We shall now give a small example. Consider the following LCP

$$j = \begin{pmatrix} 2 & 1 \\ 1 & 1 \end{pmatrix} u + \begin{pmatrix} a \\ b \end{pmatrix}$$

$$u^T j = 0, \, j, u \geq 0$$

Starting with j_1 yields

$$j_1 = 2u_1 + d_1$$
$$d_1 = a + u_2$$

and from (3.29) $j_1 = \lfloor \tilde{d}_1 \rfloor = \lfloor a + \tilde{u}_2 \rfloor$. In this example system (3.25) is expressed as $\tilde{j}_2 = \tilde{u}_2 + b$ and yielding the solution $\tilde{u}_2 = \lfloor -b \rfloor$ according to (3.22). This leads to the explicit function

$$j_1 = \lfloor a + \lfloor -b \rfloor \rfloor$$

Repeating this for the second part of j yields

$$j_2 = \left\lfloor b + \left\lfloor -\frac{a}{2} \right\rfloor \right\rfloor$$

Because

$$u = \begin{pmatrix} 1 & -1 \\ -1 & 2 \end{pmatrix} j + \begin{pmatrix} b - a \\ a - 2b \end{pmatrix}$$

we obtain the explicit functions for u

$$u_1 = \lfloor b - a - 0.5 \lfloor 2b - a \rfloor \rfloor$$
$$u_2 = \lfloor a - 2b - \lfloor a - b \rfloor \rfloor$$

The application of the proposed algorithm is limited and lies mainly in its theoretical meaning. Due to the rapid increase of the amount of work involved, only small matrices can be handled, and the complexity of the algorithm is huge.

However the discussed method is, as far as we know of, the first procedure that actually yields explicit solutions for *any class-P LCP-problem*.

To conclude two network examples will demonstrate this remarkable ability by yielding an explicit solution for a piecewise linear passive resistive network, serving the same role as would do Cramers rule for a linear passive resistive network.

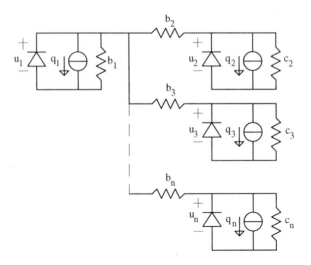

Figure 3.4. PL network with a special LCP matrix D.

The first piecewise linear network is given in Fig. 3.4 yielding the following equations

$$j_1 = q_1 + b_1 u_1 + \sum_{i=2}^{n} b_i (u_1 - u_i)$$

$$j_i = q_i + c_i u_i + b_i (u_i - u_1) \qquad \forall_{i \in \{2,...,n\}}$$

(3.30)

and where the complementary conditions should be obeyed as well. Reformulating these equations leads to the following class-P LCP

$$\begin{pmatrix} j_1 \\ j_2 \\ \vdots \\ j_n \end{pmatrix} = \begin{pmatrix} d_{11} & d_{12} & \cdots & d_{1n} \\ d_{21} & d_{22} & & \\ \vdots & & \ddots & \\ d_{n1} & & & d_{nn} \end{pmatrix} \begin{pmatrix} u_1 \\ u_2 \\ \vdots \\ u_n \end{pmatrix} + \begin{pmatrix} q_1 \\ q_2 \\ \vdots \\ q_n \end{pmatrix}$$

(3.31)

with $d_{11} = b_1 + b_2$ and $\forall_{i \in \{2,...,n\}} d_{1i} = d_{i1} = -b_i, d_{ii} = c_i + b_i$ ($d_{ij}\big|_{j \neq 1, i \neq 1, i \neq j} = 0$).

According to (3.23) we have

$$j_1 = d_{11}u_1 + \sum_{i=2}^{n} d_{1i}u_i + q_1 = d_{11}u_1 + d(u)$$

$$\hat{j} = d_{\bullet 1}u_1 + \begin{pmatrix} d_{22} & & \\ & \ddots & \\ & & d_{nn} \end{pmatrix} \hat{u} + \begin{pmatrix} q_2 \\ \vdots \\ q_n \end{pmatrix} = d_{\bullet 1}u_1 + \overline{D}\hat{u} + \overline{q} \tag{3.32}$$

$$\hat{d} = d(\hat{u})$$

and the $(n\text{-}1)$-dimensional LCP is defined as

$$\tilde{j} = \begin{pmatrix} \tilde{j}_2 \\ \vdots \\ \tilde{j}_n \end{pmatrix} = \begin{pmatrix} d_{22} & & \\ & \ddots & \\ & & d_{nn} \end{pmatrix} \begin{pmatrix} \tilde{u}_2 \\ \vdots \\ \tilde{u}_n \end{pmatrix} + \begin{pmatrix} q_2 \\ \vdots \\ q_n \end{pmatrix} = \overline{D}\tilde{u} + \overline{q} \tag{3.33}$$

The solution of (3.33) is easily obtained and given as

$$\forall_{i \in \{2,\ldots,n\}} \tilde{j}_i = \lfloor q_i \rfloor, \ \tilde{u}_i = \lfloor -q_i / d_{ii} \rfloor \tag{3.34}$$

and hence

$$\tilde{d} = d(\tilde{u}) = \sum_{i=2}^{n} d_{1i}\tilde{u}_i + q_1 = \sum_{i=2}^{n} d_{1i} \left\lfloor -\frac{q_i}{d_{ii}} \right\rfloor + q_1 \tag{3.35}$$

Because of (3.29) finally the explicit solution is obtained, yielding

$$j_1 = \left\lfloor \sum_{i=2}^{n} d_{1i} \lfloor -q_i / d_{ii} \rfloor + q_1 \right\rfloor, \ u_1 = \left\lfloor -\sum_{i=2}^{n} \frac{d_{1i}}{d_{11}} \lfloor -q_i / d_{ii} \rfloor - \frac{q_1}{d_{11}} \right\rfloor$$

$$\forall_{i \in \{2,\ldots,n\}} j_i = \lfloor q_i + d_{i1}u_1 \rfloor = \left\lfloor q_i + d_{i1} \left\lfloor -\sum_{j=2}^{n} \frac{d_{1j}}{d_{11}} \lfloor -q_j / d_{jj} \rfloor - \frac{q_1}{d_{11}} \right\rfloor \right\rfloor$$

$$\tag{3.36}$$

The solution is explicitly written in terms of values of the conductance's and current sources. Note that for this special LCP matrix the absolute-sign operator is nested up to level 3.

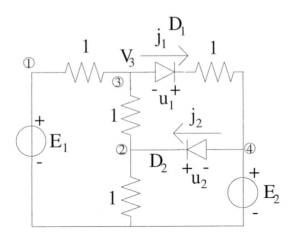

Figure 3.5. A piecewise linear network for which the explicit solution can be obtained.

As second and last example consider the piecewise linear circuit of Fig. 3.5 in which each resistor has a value of 1Ω and which contains two ideal diodes D_1 and D_2 and two voltage sources E_1 and E_2.

Taking the orientation of the diode voltages and currents as indicated in the figure, the circuit behavior is defined by

$$\begin{pmatrix} u_1 \\ u_2 \end{pmatrix} = \begin{pmatrix} \frac{5}{3} & -\frac{1}{3} \\ -\frac{1}{3} & \frac{2}{3} \end{pmatrix} \begin{pmatrix} j_1 \\ j_2 \end{pmatrix} + \begin{pmatrix} -\frac{2}{3}E_1 + E_2 \\ \frac{1}{3}E_1 - E_2 \end{pmatrix}, \text{ with } u, j \ge 0, u^t j = 0$$

Furthermore the node voltage V_3 is defined by

$$V_3 = \tfrac{1}{3}(E_1 + 2E_2 - u_1 + u_2)$$

Solving for the above LCP-problem starting with u_1 yields

$$u_1 = \left\lfloor -\tfrac{2}{3}E_1 + E_2 - \tfrac{1}{3}j_2 \right\rfloor$$

Using the second equation then yields $j_2 = \tfrac{3}{2}\left\lfloor -\tfrac{1}{3}E_1 + E_2 \right\rfloor$, which results into

$$u_1 = \left\lfloor -\tfrac{2}{3}E_1 + E_2 - \tfrac{1}{2}\left\lfloor -\tfrac{1}{3}E_1 + E_2 \right\rfloor \right\rfloor$$

Along the same lines starting with u_2 one obtains

$$u_2 = \lfloor \tfrac{1}{3}E_1 - E_2 - \tfrac{1}{5} \lfloor \tfrac{2}{3}E_1 - E_2 \rfloor \rfloor$$

Substituting the vector u into the equation for V_3 finally gives the explicit node voltage V_3 as a function of the voltages E_1 and E_2 of the voltage sources:

$$V_3 = \tfrac{1}{3}(E_1 + 2E_2 - \lfloor -\tfrac{2}{3}E_1 + E_2 - \tfrac{1}{2}\lfloor -\tfrac{1}{3}E_1 + E_2 \rfloor \rfloor + \lfloor \tfrac{1}{3}E_1 - E_2 - \tfrac{1}{5}\lfloor \tfrac{2}{3}E_1 - E_2 \rfloor \rfloor)$$

3.7. Relation to Linear Programming

In this section we will show the relation between Linear Programming (LP) and the LCP. This relation is known for some time (see for instance [37]).

To this purpose consider the primal problem:

Maximize $c^T x$

subject to

$Ax \le b$

$x \ge 0$

(3.37)

In the LP we can define for each problem its dual version. For (3.37) this will be

Minimize $b^T y$

subject to

$A^T y \ge c$

$y \ge 0$

(3.38)

Because both problems are dual to each other, in the extreme

$$c^T x = b^T y \tag{3.39}$$

must hold. Let us introduce two slack variables v and w to account for the inequality signs in (3.37) and (3.38). Then these equations become

$$Ax + v = b$$
$$A^T y - w = c \quad\quad\quad\quad (3.40)$$
$$x, v, y, w \geq 0$$

yielding

$$\begin{pmatrix} w \\ v \end{pmatrix} = \begin{pmatrix} 0 & A^T \\ -A & 0 \end{pmatrix} \begin{pmatrix} x \\ y \end{pmatrix} + \begin{pmatrix} -c \\ b \end{pmatrix} \quad\quad\quad\quad (3.41)$$

which can be treated as an LCP of the form (3.2) with $j = (w, v)^T$, $u = (x, y)^T$ and $q = (-c, b)^T$. The fact that $u^T j = 0$ holds, follows from the combination of (3.39) and (3.40) yielding

$$v^T y + w^T x = 0 \quad\quad\quad\quad (3.42)$$

For given D and q, one have to find the corresponding u and j. Obtaining u gives the solution for (3.37) and (3.38), thus minimizing $b^T y$ and maximizing $c^T x$.

3.8.   Some conclusions

In this chapter several methods to solve the LCP were discussed. Although there is not a simple answer to the question which method is the best, the Katzenelson algorithm is the most used due to its simplicity. We will see the application of this method in chapter 4 were we discuss techniques to analyze piecewise linear electronic networks. Very recently an adaptation to the Katzenelson algorithm is presented where the homotopy parameter may be complex and at least is multi parameter. The advantage is then that difficult points in the characteristic like in the hysteresis curve can be handled with more care than with the straightforward method [24].

CHAPTER **4**

PIECEWISE LINEAR ANALYSIS

The purpose of simulation of electronic networks is to obtain information on the behavior of the network for an applied source. The behavior is analyzed by computing the wave forms of the network variables, which are often the voltages and currents at the nodes of the network. In this chapter we will focus more on methods to analyze a network of piecewise linear models in an efficient way. Further we will discuss a few existing PL simulators.

4.1. Differences between simulators

Today many simulators do exist and are intensively used to analyze the behavior of a (non)linear network. Most simulators are powerful tools, able to analyze large networks in short time with a high accuracy. Simulators like SPICE are often seen as the ultimate solution kit for network analysis problems. However, many designers are not aware anymore of the limitations of the simulators they are using. First of all each simulator has a limited range of usefulness in terms of types of networks it can handle. The way the components of the network are modeled limits the operation of the simulator. Secondly, analyzing a network is in principal nothing more than solving a nonlinear function and there are many numerical aspects related to that problem. The nonlinear function determines which numerical method must be used to solve the problem and it depends on the simulator if it supports such method. Both simulation viewpoints will be discussed in more detail in the following sections.

4.1.1. The modeling viewpoint

The way the components of the network are modeled determines more or less the class of simulator. Do we consider the components on electron-hole level or are they described in terms of bandwidth and gain. The mathematical descriptions differ in both situations and determine in a strong way the numerical methods to compute the behavior of the network. Nowadays simulators can be roughly divided into five types:

- *Device Simulators*: In this class of simulators, the components are described in terms of electrons and holes. The behavior is analyzed by computing the movements of the electrons and holes through the device. This charge transport is caused by the electrical field and the differences in the concentrations. To predict the resulting charge distribution, ending up in diffusion and drift currents, one has to solve the Maxwell laws and other physical laws. The electron-hole concentrations follow the Boltzmann or Fermi-Dirac statistics. Device simulators are used to gain insight in the behavior of semiconductor devices like pn-junctions and transistor structures [41,42]. The outcome of those simulations can be used to generate models that are described in terms of external voltages and currents, to be used on the level of electrical simulations.

- *Electrical Simulators*: When the components are described in terms of voltages and currents, the electrical behavior can be computed. The components are elements like resistors, capacitors and transistors. The component equations together with the Kirchhoff laws, that describe the topology of the network, result into one nonlinear set of equations, eventually time dependent. This set of equations is then solved with simulators like SPICE [43,44], SAMSON [45] or ASTAP [46,47]. These kind of simulators are still intensively used in industry and universities and they are often referred to as circuit simulators.

- *Logic Simulators*: With the growing complexity of digital circuits with thousands of transistors, it became clear that analyzing the behavior on electrical level was almost impossible to perform with electrical simulators. Besides that, one was more interested in the functional working in terms of binary codes rather than in the exact electrical level in terms of milli volts. The demand was on the analysis of the logic functionality, which means that the simulators should be based on Boolean algebra. The outcome of such simulators is then nothing more as the truth-table of the network. Known simulators of this type are COSMOS [48] and LDSIM [49]. But because with such simulations timing problems could never be detected, switched level (SL) simulators were developed. In such simulators the transistors are modeled as ideal switches with an RC network representing the delay. Still the logic functionality could be analyzed, but now with timing information at the cost of a reduced speed. Simulators using this principle are for instance MOSSIM [50] and SLS [51].

- *Behavioral Simulators*: Sometimes networks are described in terms of mathematical operations on signals. This happens when dealing with a large complex system, for instance a complete telecommunication chip for ISDN. Before designing in detail, first the overall parameters must be settled. This can be done using behavioral simulators, which solve pure mathematical equations, without a relation to the electrical or logic level. Such a simulator is MATRIXx

[52]. However, VHDL simulators can also be seen as member of this group [53].

- *Mixed Level/Mixed Signal (MLMS) simulators*: In recent years a new branch of simulators has been developed to bridge the gap between electrical, logic and behavioral simulators. The state-of-art in VLSI demands for simulators able to analyze networks where some components are described on the electrical level while others are described in terms of gain and bandwidth. This so called mixed level simulation is of interest when not the complete circuit is designed up to transistor level but the designer still wants to know whether the circuit is functioning or not. This type of simulation is also useful for analyzing large networks, where only the functionality of small parts must be known in detail. With the possibility to integrate analog and digital circuits on a single chip, also the need for mixed signal simulators was growing. However, to gain speed, the digital part should be analyzed on binary level, instead of on electrical level. The group of MLMS simulators can be divided into two sub groups, depending on how they internally solve the network problem.

  1. *The glued approach.* In principal the MLMS simulators based on the glued approach divide the network into several parts. Each part contains components that are described in the same way and can be solved with a dedicated simulator. Thus the digital part is separated from the analog part and the behavioral part. The advantage is that for each part the best simulator can be used with its advantages in that domain. However, the network is a connection of those separated parts and thus the simulators must 'talk' to each other. The outcome of one simulator is the input for an other simulator demanding good interfacing. The main problem is in the stability of the overall simulation. Because each dedicated simulator has no overall view of the complete network, problems arise when there are feedback loops in the system with in the loop different components to be simulated by different simulators. Simulators based on the glued approach are for instance SABER [54], FIDELDO [55], SMASH [56] and to a lesser degree, later versions of SPICE.
  2. *The unified approach.* Although the network consists of components out of different classes, they are described in the same format. This allows for one solution algorithm and thus one simulator. Now it does not matter whether there are feedback loops in the network and this type of simulator can handle such networks with good convergence properties. To this class belong the piecewise linear simulators like PLATO [57], the one proposed in [58] and PLANET [59] and the latter will, as example, be discussed in more detail.

4.1.2. The numerical viewpoint

A network description can be seen as a set of nonlinear time dependent equations
for which the solutions (in time) must be obtained. This can be achieved using
numerical methods. Those numerical methods are collected into one tool kit, called
the simulator. However, each simulator consists of a small set of those numerical
methods, not necessarily enough to solve all kinds of equations and thus not able to
analyze all kinds of network problems.

A set of ordinary differential equations can be transformed into a set of time
independent equations by applying numerical integration methods. Therefore the
methods to solve a set of nonlinear time independent equations are the core of each
simulator. Those methods are used during DC analysis, but also during transient
analysis. In the latter situation, the time axis is divided into small time intervals and
in each interval the set of equations is transformed into time independent equations.
During AC analysis, when the frequency response of a network is studied, the
network is linearized around the operating point of the network. This point is
obtained as a solution of a set of nonlinear time independent equations. In a similar
way sensitivity analysis, noise analysis and other types of analysis make use of the
methods that represent the numerical core of a simulator. We will discuss the
possible methods to obtain solutions of a set of nonlinear equations as well as those
for time dependent equations.

Solving nonlinear time independent equations
To find the root of a set of nonlinear time independent equations is a large research
topic in mathematics. In the last decades many algorithms are developed, each with
their advantages and drawbacks with respect to convergence, accuracy and
computational effort. In principal the methods can be divided into a few groups,
from which two important ones are the iterative methods and the piecewise linear
methods.

From the first group, one well known, widely used, method is the Newton-
Raphson (N-R) iteration procedure. In the scalar case, the N-R iterations to solve
$f(x) = 0$ are given by

$$x_{n+1} = x_n + \Delta x_n = x_n - \frac{f(x_n)}{\dot{f}(x_n)} \tag{4.1}$$

where the subscripts denote the iteration number. The detailed theory is available in
books on numerical analysis and the reader is referred to these books for the
theoretical background. The method has a quadratic convergence property if the
initial estimate is close to the solution. However, for an arbitrary initial estimate
there is no guarantee for convergence. Practical this means that a simulator is not
always able to obtain the DC operating point of a network. Small modifications in
the network or defining an other initial node voltage set, will change the initial
estimate for the N-R procedure which ultimately give the DC operating point of the
network.

Using piecewise linear methods, the behavior of the nonlinear function is approximated with a set of piecewise linear affine mappings or segments. This result in several sets of linear mappings, as is already discussed in the previous chapters. The problem of finding the root of equations is transformed into finding which linear mapping is valid. This problem is related to the LCP, as discussed in chapter 3. The methods to solve this problem are in general less sensitive towards the initial estimate and therefore have better global convergence properties than the N-R or other iterative methods.

Using PL techniques the accuracy of the obtained solution depends on the granularly of approximation, where with iterative procedure this is determined by the convergence of the iteration sequence and, when the derivative of the function is not known, by the prediction of this derivative.

Solving ordinary differential equations
During transient analysis, the time dependent network is transformed into a sequel of time independent networks by applying numerical integration methods. In the literature many numerical integration methods are presented and discussed. From network theoretical viewpoint, the methods must be able to solve differential equations where the ratio between the time constants is large. Such equations are called *stiff* and to obtain the time solutions one needs an *A-stable* numerical integration method. Some well-known numerical integration methods are:

- *Linear Multi-Step Methods (LMS)* : This class of methods is probably the most well-known in the area of circuit simulation. The time interval over which the solution must be found is divided into intervals by selecting its boundary time steps. For every resulting time point t_n the value of x_n is calculated using a linear combination of x_j's and $\dot{x}_j$'s which can be written as

$$\sum_{j=0}^{k} \alpha_j(h)x_{n-j} + \sum_{j=0}^{k} \beta_j(h)\dot{x}_{n-j} = 0 \qquad (4.2)$$

In fact the time axis is made discrete to find an expression for the derivative of x from which the resulting algebraic system is solved using several linearization steps. Amongst the LMS formulas are the well-known Euler formulas and the trapezoidal rule as well as the Adams-Bashforth and Gear formulas [60-62]. In Fig. 4.1, the time-iteration plane for an LMS method is depicted, where it can be seen that the first guess for a new time point is the solution of the previous time point. At every time point several N-R iterations are performed to solve the resulting set of nonlinear time independent equations.

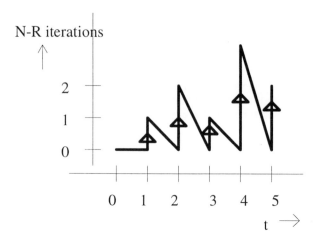

Figure 4.1. Time-iteration plane for an LMS method

The LMS methods are often extended with variable step size, where monitoring is applied to keep track on the stability error [63-65].

• *Wave form Newton Methods (WN)* : The first step in these methods is to linearize the equations around a chosen wave form. As initial guess often a constant wave form is chosen. From the resulting linear differential equation the new wave form can be computed using e.g. an LMS method. Therefore for each iteration step an approximation of the total wave form is computed, as depicted in Fig. 4.2. From a network point of view it can be shown that the nonlinear time-invariant network is repeatedly replaced by a linear time variant network. This method is still a major research topic [66-67].

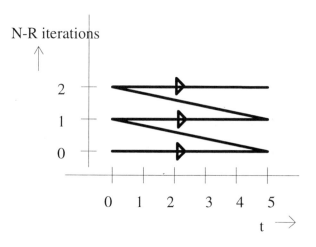

Figure 4.2. Time-iteration plane for WN method

- *One-Step Methods* : The most widely used one step methods are the class of Runge-Kutta methods (RK) [68-70], which in general for a given time step h can be given as

$$k_1 = hf(x_n, t_n)$$

$$\vdots$$

$$k_m = hf\left(x_n + \sum_{j=1}^{m-1} \beta_{m-1} k_j, t_n + \alpha_{m-1} h \right) \tag{4.3}$$

$$x_{n+1} = x_n + \sum_{i=1}^{m} \gamma_i k_i, \qquad \sum_{i=1}^{m} \gamma_i = 1$$

where the coefficients α, β, γ are chosen such that the order of the method is maximal. The most explicit RK methods are not suitable for network simulation because of the shape of the stability region. To overcome these problems implicit RK methods should be used. The advantage of RK methods is the easily implementation of variable step size methods and the large stability area versus order of the methods in implicit form.

Most simulators provide the user with the trapezoidal rule and the Gear method, which indeed are the best methods, related to the, often stiff, differential equations that must be solved. However, most simulators apply for a variable step size algorithm, which can have a seriously negative effect on the stability and accuracy of the used method.

4.1.3. Piecewise linear simulators

From the above given outline it becomes clear that PL simulators belong to the class of MLMS simulators, using a unified approach. From numerical point of view they do not rely on iterative methods to obtain the operating points, but are confronted with the LCP. However the methods to solve the LCP have stronger global convergence properties than N-R methods. PL simulators do not differ in the transient domain from other simulators and use the same kind of numerical integration methods as other types of simulators. Concluding, in PL simulators the model description is piecewise linear and the used solution algorithm is like van de Panne or Katzenelson, as discussed in chapter 3.

4.2.   Single PL model, static simulation

In chapter 2, the second model description proposed van Bokhoven was discussed and it was stated that this model could be used to advantage in a simulation environment. In this section we will demonstrate this property and to that purpose we will use a single model only. Later on we will connect models together to realize complete networks.

4.2.1. An efficient update algorithm

Let us first consider again the model description of van Bokhoven, which was defined as

$$0 = Iy + Ax + Bu + f$$
$$j = Dy + Cx + Iu + g \qquad (4.4)$$
$$u \geq 0, j \geq 0, u^T j = 0$$

with $x \in R^n, y \in R^m, u, j \in R^k$. In the following discussion we will leave out the complementary condition in the third equation for convenience.

The second equation, the state equation, defines which part of the (co) domain space is valid ánd for that part the first equation defines the linear mapping. Suppose that all u are equal zero, then the linear mapping is given by $Iy+Ax+f=0$. Crossing a hyperplane changed the signs of the state vectors and this means that a certain u entry becomes unequal zero. Although the linear mapping $Iy+Ax+Bu+f=0$ is still valid, it is eligible to have the same form of the model after the change of u has taken place. This means that the set (4.4) has to be transformed into a set like

$$0 = Iy + \bar{A} x + \bar{B} u + \bar{f}$$
$$\qquad (4.5)$$
$$j = \bar{D} y + \bar{C} x + Iu + \bar{g}$$

where the new mapping is again given by $0 = Iy + \bar{A} x + \bar{f}$ for $u=0$. This update can be done efficiently due to the fact that the state vector u in the model description is multiplied by the identity matrix. Suppose that the i-th component of u becomes positive, i.e. $u_i > 0, j_i = 0$. The expression for u_i can be obtained from the state equation in (4.4) and has to be substituted into the system equation of (4.4), which is still valid, yielding

$$
0 = Iy + Ax + B \left[-\left(D_{i1} \quad \cdots \quad D_{im} \right) y - \left(C_{i1} \quad \cdots \quad C_{in} \right) x - g_i + j_i \begin{bmatrix} u_1 \\ \vdots \\ u_{i-1} \\ u_{i+1} \\ \vdots \\ u_k \end{bmatrix} \right] + f
$$

$$(4.6)$$

Grouping related information together yields

$$
0 = \left(I - B \begin{bmatrix} 0 & \cdots & 0 \\ D_{i1} & \cdots & D_{im} \\ 0 & \cdots & 0 \end{bmatrix} \right) y + \left(A - B \begin{bmatrix} 0 & \cdots & 0 \\ C_{i1} & \cdots & C_{in} \\ 0 & \cdots & 0 \end{bmatrix} \right) x + B \begin{pmatrix} u_1 \\ \vdots \\ u_{i-1} \\ j_i \\ u_{i+1} \\ \vdots \\ u_k \end{pmatrix} + \left(f - b \begin{bmatrix} 0 \\ g_i \\ 0 \end{bmatrix} \right)
$$

$$(4.7)$$

Define $d_i = D^T e_i, c_i = C^T e_i, b_i = B e_i, g_i = g^T e_i$ and e_i the i-th Cartesian unit vector and let u_i and j_i change names, then (4.7) can be reformulated into

$$
0 = \left(I - b_i d_i^T \right)^{-1} y + \left(A - b_i c_i^T \right) x + Bu + \left(f - b_i g_i \right) \tag{4.8}
$$

where the state vector u is equal zero again. From (4.8) it can be seen that the update is dyadic, i.e. a rank one update. This was already stated in chapter 2. Using the well-known Sherman-Morrison-Woodbury relation, the inverse of $\left(I - b_i d_i^T \right)$ can be computed explicitly:

$$
\left(I - b_i d_i^T \right)^{-1} = I + \gamma_i b_i d_i^T = S_i, \quad \gamma_i = \left(1 - d_i^T b_i \right)^{-1} \tag{4.9}
$$

which for (4.5) finally yields

$$\bar{A} = S_i\left(A - b_i c_i^T\right)$$

$$\bar{B} = S_i B \qquad\qquad\qquad (4.10)$$

$$\bar{f} = S_i\left(f - b_i g_i\right)$$

The state equation has to be updated as well to obtain (4.5). Because u_i and j_i changed names, rows i of D, C and g change sign and one can easily keep track of these changes in some binary vector. We see that the update is indeed very efficiently, because mainly vector-matrix manipulations are necessary.

4.2.2. Finding the correct polytope

Suppose the model of (4.4) with $u=0$ holds for a certain input vector x_0. The question arises how to obtain the correct mapping for an input vector x_e. For the latter input vector, some entries of u in (4.4) will be unequal zero and the problem is how to obtain these. If we know them, we can update the model according to the strategy as explained above and we again obtain a valid mapping like (4.5) for x_e.

In chapter 3, the Katzenelson algorithm was explained as a solution algorithm for the LCP. Katzenelson's algorithm generated a path $x = x_0 + \lambda(x_e - x_0) = x_0 + \lambda d$ with $0 \le \lambda \le 1$ to find a sequence of u entries that where unequal zero. In the model description of (4.4) we can do the same. Substitution of the system equation into the state equation yields

$$j = (C - DA)x_0 - Df + g + \lambda(C - DA)d = v + \lambda w \ge 0 \qquad (4.11)$$

Because for a valid polytope $u=0$ and thus we have $j \ge 0$. Crossing a hyperplane means that the corresponding state vectors change role and thus j becomes zero. This can be detected by tuning λ,

$$\lambda_p = -\frac{v_p}{w_p}, w_p \neq 0, p \in \{1,\ldots,k\} \qquad (4.12)$$

where the hyperplane that is crossed first is given by the smallest λ_p larger than the previous one to ensure an adjacent polytope is entered. If $\lambda = 1$ holds, the endpoint is reached and the valid mapping for x_e is determined. In this way, the Katzenelson algorithm is reformulated as a simple check on element values. It is possible that no λ can be found in the forward search direction. This means that the search direction must be reversed and one has to search for largest λ smaller than the previous one.

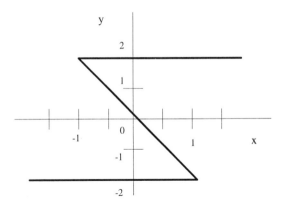

Figure 4.3. Hysteresis curve

We will demonstrate this algorithm with an example. Consider the hysteresis curve of Fig. 4.3, for which the model description is given by (see also chapter 2, section 2.3.2)

$$0 = y + (0)x + (-2 \quad 2)u + 2$$

$$j = \begin{pmatrix} -1 \\ -1 \end{pmatrix} y + \begin{pmatrix} -1 \\ -1 \end{pmatrix} x + Iu + \begin{pmatrix} -1 \\ 1 \end{pmatrix}$$

Let $x_0 = -4$, for which the mapping is already in its correct state and therefore $u=0$. The mapping for $x_e = 4$ is to be found. According to Katzenelson the path is then defined as $x = -4 + \lambda 8$. Using (4.11), one obtains

$$v = \begin{pmatrix} 5 \\ 7 \end{pmatrix}, w = \begin{pmatrix} -8 \\ -8 \end{pmatrix}$$

The smallest λ larger than zero is $\lambda_1 = \dfrac{5}{8}$ with the first entry of u which changes sign. This λ corresponds with $x=1$, where indeed the boundary between two adjacent   segments   is   located.   Using   relation   (4.10)   and $b_1 = -2, c_1^T = -1, d_1^T = -1$ and $g_1 = -1$, the update of the model description is given as

$$0 = y + (2)x + (2 \quad -2)u + 0$$

$$j = \begin{pmatrix} 1 \\ -1 \end{pmatrix} y + \begin{pmatrix} 1 \\ -1 \end{pmatrix} x + Iu + \begin{pmatrix} 1 \\ 1 \end{pmatrix}$$

The vectors v and w are then computed as

$$v = \begin{pmatrix} 5 \\ -3 \end{pmatrix}, w = \begin{pmatrix} -8 \\ 8 \end{pmatrix}$$

From this point it appears that there is no smallest λ greater than λ_1 and therefore one searches in opposite direction. This is also clear from Fig. 4.3, to enter the middle segment, the direction in x must be reversed. We want to perform this step backwards as small as possible, because our goal is to reach the point for $\lambda=1$. We find now $\lambda_2 = \frac{3}{8}$, where $x=-1$, the other boundary of two polytopes. The new update of the model is then given as

$$0 = y + (0)x + (-2 \quad 2)u + (-2)$$
$$j = \begin{pmatrix} 1 \\ 1 \end{pmatrix} y + \begin{pmatrix} 1 \\ 1 \end{pmatrix} x + Iu + \begin{pmatrix} 1 \\ -1 \end{pmatrix}$$

with corresponding vectors

$$v = \begin{pmatrix} -1 \\ -3 \end{pmatrix}, w = \begin{pmatrix} 8 \\ 8 \end{pmatrix}$$

Now it is possible to increase λ to 1 without j getting a negative entry. The polytope for x_e is reached and the valid mapping is $y-2=0$, which is in agreement with Fig. 4.3.

From the example it can be seen that with the Katzenelson algorithm and the updating procedure for a given input vector x of a PL model, the corresponding output vector can be obtained.

4.2.3. The algorithm

The method given above can be summarized in the following pseudo-algorithm that calculates y for a given x_e where a starting point x_0 is known in the polytope in which the model has been initialized. Vector s is a binary vector with entries taken from $\{-1,1\}$ to account for the signs of the rows of D, C and g.

---

begin createvw(v,w)
    $v = (C - DA)x - Df + g$
    $w = (C - DA)d$
end createvw

begin find_pivot(v,w, λ,pivot,previous_pivot)
    λ=1,pivot=0
    for $\forall_{i\in\{1...k\}}$ **do**   {check all pivots}

        if $w_i < 0$ **then** $\lambda_i = -v_i/w_i$ {try to increase λ}

        if $\lambda_i < \lambda$ **then** ($\lambda = \lambda_i$, pivot=i)

    endfor   {find minimum λ}
    if pivot=previous_pivot **then**   {no step forward possible}
        λ=0
        for $\forall_{i\in\{1...k\}}$ **do**   {check all pivots}

            if $w_i > 0$ **then** $\lambda_i = -v_i/w_i$ {try to decrease λ}

            if $\lambda_i > \lambda$ **then** ($\lambda = \lambda_i$, pivot=i)   {find maximum λ}
        endfor
    endif
    previous_pivot=pivot
end find_pivot

begin update(i)   {update system equations}

$$c_i^T = s(i)C_{i\bullet}$$

$$d_i^T = s(i)D_{i\bullet}$$

$$g_i = s(i)g_i$$

$$\bar{A} = S_i\left(A - b_i c_i^T\right)$$

$$\bar{B} = S_i B$$

$$\bar{f} = S_i\left(f - b_i g_i\right)$$

end update

begin main(x_e ,system equations)

    for $\forall_{i\in\{1...k\}}$ **do** s(i)=1

        createvw(v,w)
        previous_pivot=0
        find_pivot(v,w, λ ,pivot,previous_pivot)
        while pivot ≠0 **do**
            update(pivot)
            createvw(v,w)
            find_pivot(v,w, λ,pivot,previous_pivot)
        endwhile
    $y = -Ax_e - f$
end main

4.3. Single PL model, transient simulation

As explained in the beginning of this chapter, a dynamic model can be transformed into a time independent model by applying numerical integration methods. The use of such methods can be seen as electrically equivalent to the replacement of all capacitors by a series network of a resistor with a voltage source, where the element values depend on the particular integration formula. Consider for instance the trapezoidal rule, given as $x_{n+1} = \frac{h}{2}(\dot{x}_{n+1} + \dot{x}_n) + x_n$, applied to a (linear) capacitor. The voltage across the capacitor and the current flowing through it are related by $i = c\dfrac{dv}{dt}$. Identify the voltage v as x then this yields

$$v_{n+1} = \frac{h}{2}(\frac{i_{n+1}}{C} + \frac{i_n}{C}) + v_n = \frac{h}{2C}i_{n+1} + (\frac{h}{2C}i_n + v_n) = Ri_{n+1} + E_n$$

that indeed yields a resistor in series with a voltage source.

Consider again the general form of a k-step variable step size linear multi step integration formula (4.2) that can be rewritten into

$$\sum_{j=0}^{k}\alpha_j(h)x_{n-j} + \sum_{j=0}^{k}\beta_j(h)\dot{x}_{n-j} = 0$$

(4.13)

$$x_{n+1} = \sum_{j=0}^{k-1}\alpha_j(h)x_{n-j} + \sum_{j=-1}^{k-1}\beta_j(h)\dot{x}_{n-j} = P_n + \beta_{-1}(h)\dot{x}_{n+1}$$

where h is the current step size. The dynamic model description can be given as (see chapter 5 for more details)

$$0 = Iy + Ax + Bu + Hz + J\,\dot{z} + f$$

$$0 = Ky + Lx + Mz + N\,\dot{z} + r \qquad (4.14)$$

$$j = Dy + Cx + Iu + g$$

and eliminating $\dot{z}$ using (4.13) leads to the static model

$$0 = I\begin{pmatrix} y_{n+1} \\ z_{n+1} \end{pmatrix} + \begin{pmatrix} A_{11} & A_{12} \\ A_{21} & A_{22} \end{pmatrix}\begin{pmatrix} x_{n+1} \\ P_n \end{pmatrix} + \begin{pmatrix} B_1 \\ B_2 \end{pmatrix}u + \begin{pmatrix} f_1 \\ f_2 \end{pmatrix}$$

$$j = \begin{pmatrix} D & D_2 \end{pmatrix}\begin{pmatrix} y_{n+1} \\ z_{n+1} \end{pmatrix} + \begin{pmatrix} C & C_2 \end{pmatrix}\begin{pmatrix} x_{n+1} \\ P_n \end{pmatrix} + Iu + g \qquad (4.15)$$

with

$$A_{11} = A + \psi\theta^{-1}(L - KA)$$
$$A_{12} = (KJ^* - N^*) - J^*$$
$$A_{21} = \theta^{-1}(L - KA)$$
$$A_{22} = \theta^{-1}(KJ^* - N^*)$$
$$B_1 = B - \psi\theta^{-1}KB$$
$$B_2 = -\theta^{-1}KB$$
$$f_1 = \psi\theta^{-1}(r - Kf) + f$$
$$f_2 = \theta^{-1}(r - Kf)$$
$$N^* = \beta_{-1}^{-1}(h)N$$
$$J^* = \beta_{-1}^{-1}(h)J$$
$$\theta = (M - KH + N^* - KJ^*)$$
$$\psi = -(H + J^*)$$
$$D_2 = 0$$
$$C_2 = 0$$

Further one can prove that the following statement holds:

$$P_{n+1} = \bar{P}_n + \alpha_0 z_{n+1} + \frac{\beta_0(h)}{\beta_{-1}(h)}(z_{n+1} - P_n)$$

$$\bar{P}_n = \sum_{j=0}^{k-1} \alpha_{j+1}(h) z_{n-j} + \sum_{j=0}^{k-1} \beta_{j+1}(h)\dot{z}_{n-j} \tag{4.16}$$

One can see that (4.15) has the same format as (4.4) and therefore the same pivoting algorithm can be used. To obtain the correct pivot, still the method of Katzenelson can be applied, however, the excitation vector can now be described as

$$\begin{pmatrix} x_n(\lambda) \\ P_{n-1}(\lambda) \end{pmatrix} = \begin{pmatrix} x_n \\ P_{n-1} \end{pmatrix} + \lambda \left[\begin{pmatrix} x_{n+1} \\ P_n \end{pmatrix} - \begin{pmatrix} x_n \\ P_{n-1} \end{pmatrix} \right] \tag{4.17}$$

In the same way as in the static situation vectors v and w can be defined to obtain the correct pivots. For a fixed step size algorithm the solution strategy is the same for each time step and the above outlined method can therefore be summarized in the following algorithm.

---

begin one_timestep
   read($x(\lambda)$)   {read x_{n+1}, $x(\lambda) = x_n + \lambda(x_{n+1} - x_n)$}
   calc_P_n(integration_method)   {compute (4.16)}
   createvw(v,w)   {use **createvw**}
   λ=0
   previous_pivot=0
   find_pivot(v,w, λ ,pivot,previous_pivot)   {use **find_pivot**}
                                    {solve system}

   while(λ<1) **do**
      update(pivot)
      createvw(v,w)
      find_pivot(v,w, λ,pivot,previous_pivot)
   endwhile
end one_timestep

The above outlined method can be extended with variable step size and the corresponding control mechanism, but the main flow will still hold.

4.4.   Single PL model, other simulation domains

Besides static analysis and transient analysis, most simulators provide the user also
with the possibility of other kinds of analysis like AC, harmonic distortion or
sensitivity analysis. Because those simulations do not rely on the PL principles, we
will briefly discuss those analysis forms.

AC simulation
In AC simulation, the small signal behavior at a certain DC bias point is considered.
This means that the system is considered in exactly one polytope, and thus for this
polytope becomes $u = 0$. The proper system equations then become

$$0 = Iy + Ax + Hz + J\dot{z} + f$$

$$0 = Ky + Lx + Mz + N\dot{z} + r \tag{4.18}$$

that, using the Laplace transform can be written as

$$0 = I\bar{y} + A\bar{x} + (H + sJ)\bar{z}$$

$$0 = K\bar{y} + L\bar{x} + (M + sN)\bar{z} \tag{4.19}$$

where $\bar{a}$ denotes the Laplace transform of a and s the Laplace variable. The vectors
f and g are omitted because they play no further role in AC analysis. Eliminating the
state variable and separating the real and imaginary parts leads to

$$0 = \begin{bmatrix} I & 0 \\ 0 & I \end{bmatrix} \begin{bmatrix} y_r \\ y_j \end{bmatrix}$$

$$+ \left\{ \begin{bmatrix} A & 0 \\ 0 & A \end{bmatrix} - \begin{bmatrix} H & - \\ & H \end{bmatrix} \begin{bmatrix} M\text{-}KH & - \\ (N\text{-}KJ & M\text{-}KH \end{bmatrix}^{-1} \begin{bmatrix} L\text{-}KA & 0 \\ 0 & L\text{-}KA \end{bmatrix} \right\} \begin{bmatrix} x_r \\ x_j \end{bmatrix} \tag{4.20}$$

where ω is the radial parameter and the subscripts r and j represent the real and
imaginary part of the parameters respectively. This formula is according to $Iy + Ax$
$= 0$ where A is now a function of the frequency. For each frequency one must solve
a set of linear equations.

Harmonic distortion
To analyze the harmonic distortion (HD) of an output signal compared to that of the
input signal a transient analysis has to be performed. The input signal is normally a
sine wave form with a fixed frequency and amplitude. The output signal will be
transformed into the Fourier domain. The second order harmonic distortion is then
defined as the ratio of the amplitude of the frequency component at twice the
nominal frequency and the nominal frequency. The nominal frequency is defined as
the frequency of the input signal.

Sensitivity analysis
In sensitivity analysis the change in the behavior of the circuit due to small
variations in the model parameters is determined. Maybe the most obvious way to
determine sensitivity is to do several simulations with slightly different parameter
values. A far more elegant method is to use the *adjoint network approach* [71-73].
In the adjoint network approach we deal with two networks: First there is the
network under consideration i.e. the network of which we want to determine the
sensitivities, which is linearized in some chosen bias point. The resulting linear
network will be called *the normal network*. Secondly there is *the adjoint network*
which is topological identical to the normal network. However, every component is
replaced by its adjoint counterpart: suppose that a component is characterized by a
hybrid matrix like

$$\begin{pmatrix} A_{11} & A_{12} \\ A_{21} & A_{22} \end{pmatrix} \begin{pmatrix} v_1 \\ i_2 \end{pmatrix} = \begin{pmatrix} i_1 \\ v_2 \end{pmatrix} \tag{4.21}$$

then the adjoint component is described by

$$\begin{pmatrix} A_{11}^T & -A_{21}^T \\ -A_{12}^T & A_{22}^T \end{pmatrix} \begin{pmatrix} \tilde{v_1} \\ \tilde{i_2} \end{pmatrix} = \begin{pmatrix} \tilde{i_1} \\ \tilde{v_2} \end{pmatrix} \tag{4.22}$$

where the superscript T denotes transposed and the tilde denotes the network
variables in the adjoint network. The transformation from (4.21) into (4.22) comes
down to transposing the total matrix and adding a minus sign to the entries that are
dimensionless. The sensitivity of some voltage with respect to a change in
parameters can then directly be obtained from two simulations, one with the normal
network and one with the adjoint network.

4.5. Connecting PL models, the complete network

In the section 4.2 a method was presented to update the component equations in an efficient way. The method described, applies to single components only. This section presents a method to simulate circuits that consist of more than one component. Further it will be explained how the hierarchy in a system can be preserved. This is achieved using several sets of topological equations on different levels and keeping the model equations separate from the topological equations.

4.5.1. One level of hierarchy

To describe the models, again (4.4) is used. Several of these models are interconnected to make up a circuit. Topological relations are imposed and added using the Kirchhoff voltage and current laws. This is shown in Fig. 4.4, where the ellipse denotes the set of topological equations. In this tree, the models are called leafs, the connection between them a node. Once this is done, the PL model descriptions and the topological equations define the circuit. The circuit can be stored in an overall system matrix depicted in Fig. 4.5. The upper part of the system matrix can be rearranged by exchanging rows and columns to form one large PL-model.

For convenience, the separate input vectors x are grouped together according to (4.23) for a circuit having n components

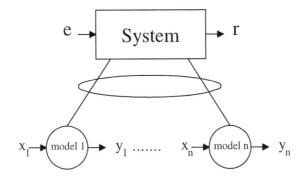

Figure 4.4. Connecting several leafs, one level of hierarchy

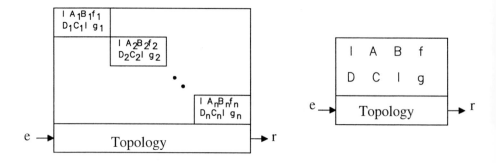

Figure 4.5. The overall system (left) and its transform (right)

$$x^T = (x_1^T, x_2^T, \ldots, x_n^T)$$ (4.23)

The same method is used for the output vector y and the state vectors u and j,

$$y^T = (y_1^T, y_2^T, \ldots, y_n^T)$$
$$u^T = (u_1^T, u_2^T, \ldots, u_n^T)$$ (4.24)
$$j^T = (j_1^T, j_2^T, \ldots, j_n^T)$$

A comparable division is used for the matrices resulting in block diagonal matrices. In (4.25) the matrix A is taken as an example

$$A = \begin{bmatrix} A_1 & & & \\ & A_2 & & \\ & & \ddots & \\ & & & A_n \end{bmatrix} \quad \text{while } y = -Ax - f$$ (4.25)

The circuit given by (4.25) is excited by a number of signals represented by the vector e. This vector e allows to define the topological equations in a form like

$$P_c x - Q_c y = R_c e$$ (4.26)

with P_c, Q_c and R_c matrices of appropriate dimensions having entries taken from $\{0,1,-1\}$ only. These in fact represent the Kirchhoff voltage and current laws. Using (4.25) it is possible to eliminate y from (4.26) and solve the independent input vector x as a function of the excitation vector e

$$P_c x - Q_c(-Ax) = R_c e \Rightarrow T_c x = R_c e \text{ with } T_c = P_c + Q_c A \qquad (4.27)$$

The constant vector f is omitted for convenience. A similar form as (4.26) can be used to define the output vector r

$$r = P_r x - Q_r y + R_r e$$
$$r = T_r x + R_r e \text{ with } T_r = P_r + Q_r A \qquad (4.28)$$

It is now possible to determine x (and consequently y) for a given e as the solution of the linear set of equations (4.27), for instance by Gaussian elimination. Knowing this vector x it is possible to find a next pivot applying Katzenelson's algorithm. The vector r can be calculated by means of (4.28). It is easily seen that by elimination x from (4.26) and (4.28) these together define a linear mapping from e to r. In a closed form expression this mapping could be denoted by A^*

$$r = A^* e \qquad (4.29)$$

In this matrix A^* the topological relations are included and normally this system is solved by simulators like SPICE. Note that in our situation however relation (4.27) is to be solved and not (4.29). When changing e from e_o to e_e as a result of the solution algorithm pivots will occur which means that one of the leafs (i.e. a component of the network), say k, has to be updated by means of the method outlined in the previous section. The update of A_k is a dyadic product only as shown in previous sections. From (4.25) it can then be seen that the update of A is also a dyadic update. This means that the updated matrix A may be written as

$$\overline{A} = A + \alpha\beta^T \qquad (4.30)$$

This update of A must also be used in (4.27) and (4.28) to get the correct x and r. The dyadic update of A results in a dyadic update of T_c and T_r.

It should be emphasized at this point that in a practical realization there is no huge system matrix and the matrices like in (4.25) do not exist: there is a collection of models and a set of topological equations. If a pivot is found, only the model in which the pivot occurs is updated.

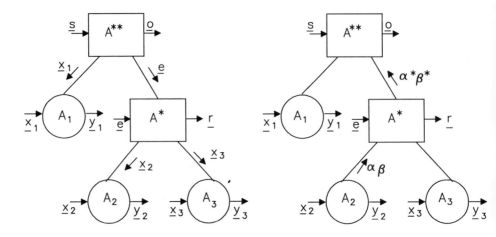

Figure 4.6. Hierarchical simulation: determination of excitation vectors (left) and update process (right)

4.5.2. Hierarchy

To be able to govern several levels of hierarchy the method outlined above is applied iteratively. The linear mapping e to r is given by (4.29) as a function of the linear mappings A_k of the components one level down in hierarchy with respect to e (and r) and the topological relations. For a given e the vector x can be calculated using (4.27). If one of the mappings of the components A_k changes by a dyadic product, the mapping A^* from e to r has to be updated also. Going up one level in the hierarchy with respect to e, several blocks with mappings A_i^* are connected. This process of updating A and computing x is depicted in Fig. 4.6. To use the method outlined above it is necessary to express the update of A^* as a dyadic product also. It is indeed possible to proof that the update of A^* is given by [59]

$$\overline{A^*} = A^* + \alpha^* \cdot \beta^{*T} \tag{4.31}$$

where α^* and β^{*T} are related simple to α, β^T, T_c and T_r.

Now that α^* and β^{*T} are calculated, the same method can be used at one level higher in hierarchy by connecting several blocks on this current level.

It should now be clear that if a pivot is found in a leaf cell, this cell must be updated for which the efficient algorithm of the previous section can be used. Also the expressions (4.26) and (4.28) of every node in the hierarchical tree from this leaf cell to the root have to be updated. The next step is to determine the excitation vectors e on every level top-down. If the excitation vectors on a certain level are known, the excitation vectors one level down in hierarchy can be found using (4.26). Because the sources exciting the entire network are prescribed, all excitation vectors e can be calculated iteratively. If the bottom level is reached there is a prescription for every vector x for every leaf cell at which new pivots can be found. It can also be seen that replacing a node by a single leaf or vice versa is facile because a leaf cell as well as a compound model is characterized by a linear mapping.

4.5.3. Finding the DC operating point

The components that make up a network are individually initialized within a certain polytope. For that particular polytope the state vector u is equal zero. A valid starting point, which lies inside the polytope, can be associated with x_{init}. If the models are interconnected as they are initialized, the Kirchhoff voltage and current laws will be violated. This because it is not necessarily true that for all the components x_{init} are the same. In other words, we have to find the overall x_{init} that can be used as starting point for the algorithm of Katzenelson. This starting point is in the network theory called the DC bias point or operating point of the network and the following steps have to be taken on order to find it.

First all dynamic element models have to be removed and replaced by an appropriate circuit (a capacitor is replaced by an open circuit and a coil by a short circuit). To abolish the violation of the Kirchhoff laws, artificial sources have to be added. This can be seen in the topological equations (4.26) that become

$$T_c x = R_c e_o + e_{art} \tag{4.32}$$

where e_o is the starting point of the excitation vector and e_{art} are the artificial sources. To find the DC bias point for an input vector e_o, the artificial sources are turned to zero using the Katzenelson algorithm according to

$$e_{art}(\lambda) = e_{art} - \lambda e_{art} \tag{4.33}$$

If λ moves from 0 to 1, pivots will occur which results in subsequent updates of T_c and a new values for $x(\lambda)$. When $\lambda=1$, (4.33) is valid for $e_{art} = 0$, the correct polytopes for the components are found and $T_c x = R_c e_o$ will hold. This means that the sub network has as starting point e_o. In the algorithm of Katzenelson this point can be treated the same way as a component model from which x_{init} is known. In this way it is possible to treat the hierarchy tree bottom up until at the root the DC bias point for the whole network is obtained. Now that the capacitor voltages and coil currents are known, the capacitors and coils are reinserted.

4.5.4. The algorithm for DC simulation

The basic purpose of simulation is to find the output signals o of a network if it is excited with prescribed sources s (see also Fig. 4.6). The first step in simulation is setting up the (hierarchical) structure, meaning that all matrices of the nodes and component models are calculated or read in. Because the components are arbitrary biased, the Kirchhoff voltage and current laws will be violated and therefore the next step is to calculate the DC bias or operating point. Once this is done, all the mappings of the nodes and components represent a linear network. For a given input $s = s_0 + \lambda(s_e - s_0)$, the excitations for all nodes and leafs have to be computed. Knowing these, it is possible to find the component where an adjacent polytope is entered. Using this information, the network can upwards be updated using dyadic vector multiplication's. A new s can be taken which is followed by the same sequence of mentioned steps. This can be summarized in the following pseudo code.

```
begin dc_simulation
     setup_tree()                                    {parse the network}
     find_dc_biaspoint()                             {solve (4.39-4.40)}
     solve_e()                              {solve (4.34) for each component}
     while(not end of sweep interval)do
```

$$\textbf{while}(\text{no change in } s = s_0 + \lambda(s_e - s_0))\textbf{do}$$

```
          find_pivot()       {find pivot over all components using find_pivot}
          update_path()       {use update for component and (4.38) for path}
          solve_e()                              {update all excitations e,x}
     endwhile
     solve_e()
     endwhile
end
```

4.5.5. Remarks

Topology and hierarchy for dynamic components can be treated in the same way as for static components. In section 4.3, we already showed that a dynamic model will be transformed into a static model using an LMS algorithm. This static model is excited by two sources. An independent source vector x resulting from the sources outside the component and an independent source vector P_n, resulting from the companion models. One can easily prove that a dynamic sub network can be treated in the same way as a single dynamic component [59]. Therefore, hierarchical oriented simulation with dynamic components can be treated in the same way as with static components.

The same holds for AC simulations. After the operating point of the network is obtained, the dynamic components are transformed according (4.20) and the network becomes frequency dependent.

It must be mentioned that hierarchical simulation has, in principle, nothing to do with mixed level simulation. Mixed level simulation means that the behaviors of components are described on different levels of abstraction. Hierarchical simulation means that some components are grouped together and internally treated as a sub network. These components can all be transistors but can also be described on mixed levels.

Obviously, the above outlined methods as implemented in the piecewise linear simulator PLANET [59] are one possible example of implementing techniques to analyze piecewise linear circuits. In the following sections will discuss briefly other possibilities.

4.6. The program PETS

The program Power Electronics Transient Simulator (PETS) applies integration techniques like discussed before on network differential equations and uses a modified Katzenelson algorithm to solve piecewise linear equations [91]. In this circuit the PL models are defined for each state σ by there boundary conditions

$$c^{\sigma} = C^{\sigma} y + c_0^{\sigma} \geq 0 \tag{4.34}$$

with y representing the output vector. For each state the linear mapping is defined by

$$H^{\sigma} y = Px \tag{4.35}$$

in which H represents the system matrix and P is the source incidence matrix, dependent on the topology but not dependent on the network state. Note that in this simulator all states and mappings are tabled and not stored in a compact closed manner. The advantage will become clear in case of discontinuities within the mapping, but the drawback is that for large systems the data storage is huge.

It is assumed that system (4.35) has a unique solution for all possible states of the network. Suppose now that the network solution y_{old} and the corresponding state σ_{old} with all boundary conditions (4.34) satisfied, are known for some vector of values of independent sources $x_{G,old}$. Here we assume that the source vector is partitioned as $x = (x_G, x_P(\sigma))^T$ where the second part contains the right hand side parameters in the characteristic of the PL elements for the given state σ. Also a new vector $x_{G,new}$ is given. The problem is to find the new solution y_{new} of the network.

One simply can generate a path by means of Katzenelson according to

$$x = x_{old} + \lambda(x_{new} - x_{old}), \qquad 0 \le \lambda \le 1 \tag{4.36}$$

having as result that the output vector and the vector of boundary conditions change linearly,

$$y = y_{old} + \lambda(y_{new} - y_{old}) \tag{4.37}$$

$$c^{\sigma} = c_{old}^{\sigma} + \lambda(c_{new}^{\sigma} - c_{old}^{\sigma}) \tag{4.38}$$

The Katzenelson parameter is found as

$$\lambda_{CS} = \min \left\{ \begin{array}{l} \dfrac{c_{i,old}^{\sigma}}{c_{i,old}^{\sigma} - c_{i,new}^{\sigma}} \text{ for } c_{i,new}^{\sigma} < 0 \\[2ex] 1 \text{ for } c_{i,old}^{\sigma} \ge 0 \end{array} \right\} \tag{4.39}$$

So far the method is the same as treated before. However, due to discontinuities, the output vector $y(\lambda_{CS}^{+})$ immediately after the state change may not be the same as the solution $y(\lambda_{CS}^{-})$ before the state change. Therefore (4.37) is violated. To solve this problem the program relies on the fact that it has all mappings available explicitly. To pass the discontinuity an additional path in the new state is created

$$c^{\sigma} = c_{old}^{\sigma} + \mu(c_{new}^{\sigma} - c_{old}^{\sigma}) \tag{4.40}$$

Because this new state is not correct in the sense that it does not contain a solution, at least one state change should occur for $0 \le \mu \le 1$. The corresponding μ can be found using a similar expression like (4.39). The obtained network solution is used as new initial point y_{old} and the newly obtained state of the network is the new guess of the final state. The algorithm can now proceed until the solution for μ equals one is obtained. Then the main iteration loop over λ may be proceeded.

The iteration over μ may become an infinite loop if the same network state is repeated. This occurs when the network does not have a solution for the given input vector. The loop must then be terminated.

The advantage of the algorithm is clear but demands that at each time for a given state the linear mapping is known. Otherwise vector y can not be computed. In case of a closed implicit model description the proposed technique is not possible. For the given state, the discontinuity is detected by the fact that γ_i^{-1} in (4.9) becomes zero and hence no update of the model is possible. Therefore we do not know the linear mapping in the new state and we loose control over the model. Too see this consider for instance the model of an ideal step function,

$$y + (0)x + \begin{pmatrix} 1 & -1 \end{pmatrix} u + 1 = 0$$

$$\begin{pmatrix} 1 \\ 1 \end{pmatrix} y + \begin{pmatrix} 1 \\ 1 \end{pmatrix} x + Iu + \begin{pmatrix} 1 \\ -1 \end{pmatrix} = j \qquad (4.41)$$

with the characteristic as depicted in Fig. 4.7. Coming from the positive side of x we have the second row in the state equation as pivot. This means that $d_1 b_1 = 1 \cdot 1 = 1$ and hence $\gamma^{-1} = (1 - d_1 b_1) = 0$. We are not able to update the model. The linear map in the middle region should read $(0)y + x = 0$ but because from analysis point of view we demand always the output vector to be known, which is in this model not the case. So, although we are able to model a discontinuity we can not handle the model. A possible strategy to overcome this problem is to interchange the role of x and y at the discontinuity.

As mentioned, in PETS the price to be paid is the data storage of the PL model, each mapping must be known in each state.

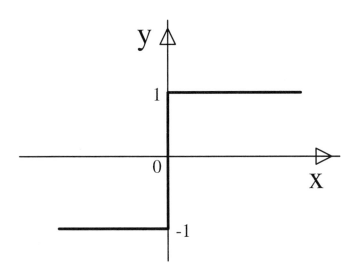

Figure 4.7. Characteristic of the model (4.41), an ideal step function

4.7. NECTAR2

Probably the oldest piecewise linear simulator is the program NECTAR2, a circuit analysis program based on piecewise linear approach [92].

The simulator makes use of the generalized Katzenelson algorithm as presented in chapter 3. The set of the system equations is described according to

$$y = PH(x) + Qx \qquad (4.42)$$

where P and Q are a pair of matrices of dimension R^{nxn} representing the interconnections of the network and H represents the characteristics of the elements in a hybrid form. Under this assumption, the pair (P, Q) has the following property

$$Pz + Qx = 0 \Rightarrow \langle z, x \rangle = 0 \qquad (4.43)$$

which is an alternative form of the Tellegen's theorem.

In this case H is piecewise linear and provided that one can start in a point, not lying on a boundary hyperplane, the Katzenelson algorithm will always obtain a solution for (4.42) in a finite number of steps.

NECTAR2 uses the backward Euler formula for numerical integration. This because of the simplicity and the naturally strong damping property of the method against parasitic effects. To improve the accuracy against truncation errors the Romberg extrapolation scheme was combined with it.

To handle large systems, a sparse tableau formulation was implemented also, where different tableaus were used for DC analysis and transient analysis.

4.8. PLATO

The Piecewise Linear Analysis Tool PLATO is an other circuit simulator developed at the Technical University of Eindhoven [16]. In contrast to PLANET this simulator uses the most general piecewise linear model description *Bokh1*, to allow a larger class of functions to be handled. Because the LCP matrix does not have the special property of being the identity matrix, hierarchy as exploit in PLANET is not possible. To overcome this problem, the developers implemented powerful methods to exploit the sparsity of the matrices and to retain this sparsity during pivoting operations. Furthermore PLATO makes use of the van de Panne algorithm to solve the LCP.

Because of the combination of the solution algorithm and the choice of model description, ideal step function can easily handled within this simulator. Tests are done to determine if in case of parallelization of the algorithms faster computations can be achieved. Although this seems to be the case provided the additional overhead costs are sufficiently low, the gain is in most situations not large enough to be acceptable.

4.9. Concluding remarks

In this chapter we discussed several issues related to analyze a network consisting of piecewise linear components. We also gave in pseudo language the algorithms to perform the analysis tasks. Finally we treated the aspect of retaining hierarchy of the network in the analysis structure.

A hierarchically structured (PL) simulator has some distinct advantages over simulators that do not retain the hierarchy of the network:

- It is easy to replace parts of the network by other networks, higher level components or vice versa. Because the data information of a single component like (4.4) is the same as for a complete network like (4.29) and both are updated by a dyadic product only, the replacement of the one in the other will not affect the simulation structure.
- To speed up computation during transient analysis, numerical integration methods with variable step size are used to exploit latency in the circuit. In the hierarchical structure it is relative simple to detect which parts of the network are dormant and for these parts not a complete integration step has to be performed but a simple extrapolation will do.
- Because the component models are separated from the topological equations, it is possible to use the efficient update algorithm because the unit matrix in the model description does not suffer from fill-in as would be the case when the models were not kept apart.

To be complete a few other PL simulators were discussed also.

CHAPTER **5**

PIECEWISE LINEAR MODELING OF ELECTRICAL CIRCUITS

In the previous chapter it was stated that the piecewise linear simulators belong to the group of MLMS simulators based on the unified approach. This means that all components have the same data format. In this chapter we will demonstrate the feasibility of the PL concepts towards component modeling in practical network situations. For that purpose we will use the format proposed by van Bokhoven, i.e. Bok2 as treated in chapter2. For the sake of clearness we will leave out the linear complementary restrictions on the state vectors u and j.

5.1. Component modeling

In this section we will discuss some examples showing the possibility to model several types of components, i.e. logic, electrical and behavioral components. However, first of all we shall shortly introduce the approximation theory, the basis of piecewise linear modeling.

5.1.1. Approximation theory

Piecewise linear techniques rely on the approximation of a nonlinear function with a set of linear function. To obtain the best set of linear functions, one can use the theorems developed in the approximation theory in the mathematics. For PL approximation the Chebychev approximation theory can be used, which is based on the important alternation theorem [1].

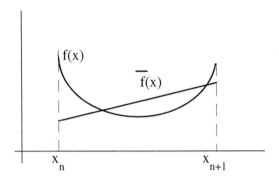

Figure 5.1. Function approximation, alternation theorem

Suppose function $f(x)$ must be approximated with an maximum error $f(x) - \overline{f}(x) = \delta$ on each interval $x_{n+1} - x_n = \varepsilon$ and with $\overline{f}(x) = a_n x + b_n$ the approximation function on that interval. The alternation theorem states that for a linear approximation $f(x)$ must alternate twice around the approximated function in each interval, as depicted in Fig. 5.1.

For $f(x) = x^2$ this yields

$$a_n = 2x_n + \varepsilon, b_{n+1} = \delta - \frac{a_n^2}{4}, \varepsilon = \sqrt{8\delta}$$

for a given error δ. For the first interval we obtain

$$a_0 = 2\sqrt{\delta}, x_1 = 0.5(a_0 + \varepsilon)$$

and further we have $x_{n+1} = x_n + \varepsilon$. In this way the linear segments can be obtained. For $\delta = 0.1$ this yields in two digit accuracy

$$\varepsilon = 0.89 \quad a_0 = 0.63 \quad x_1 = 0.76$$
$$a_{n+1} - a_n = 1.78 \quad x_{n+1} - x_n = 0.89$$

For scalar functions this routine can be automated with as output the PL model description. However, multi-dimensional functions are a more difficult task. We will discuss this topic in section 5.6. In the following sections we will demonstrate that for the most components of a network the models can be obtained by hand with not too much effort.

5.1.2. Logic components

Modeling of logic components in PL technique is done using threshold logic. The concept of threshold logic will be discussed in detail in section 5.2, but the basic idea is that the output can be treated as a weighted sum of the binary inputs. For an AND-gate this looks like Fig 5.2 where it can be seen that two hyperplanes are necessary to define three line segments. The location of the two hyperplanes can be seen in the logic diagram (Fig. 5.2). The planes are located in such way that the logic output zero's are separated from the logic output ones. This separation could be done with only one line (i.e. one hyperplane), but then resulting in a discontinuity in the threshold function. Therefore two hyperplanes are used, resulting in

$$x_1 + x_2 = \frac{5}{4}, x_1 + x_2 = \frac{7}{4}$$

The slope of the segment in the transition area is equal to 2. Grouping together this results in the following model description, valid for binary inputs,

$$0 = y + \begin{pmatrix} 0 & 0 \end{pmatrix} x + \begin{pmatrix} -2 & 2 \end{pmatrix} u + \begin{pmatrix} 0 \end{pmatrix}$$

$$j = \begin{pmatrix} -1 & -1 \\ -1 & -1 \end{pmatrix} x + Iu + \begin{pmatrix} \frac{5}{4} \\ \frac{7}{4} \end{pmatrix} \quad (5.1)$$

In this way all kinds of linear separable logic functions can be modeled. In section 5.2 we will come back to this issue.

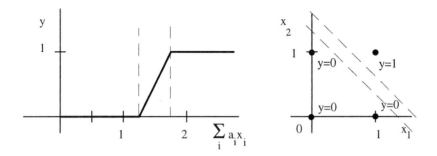

Figure 5.2. AND-gate with the threshold function (left) and the logic diagram (right)

5.1.3. Electrical components

The modeling of components like resistors and transistors can also be done in the piecewise linear model description. The model of a linear resistor for instance is relative simple, because there is a linear relation between the current through and voltage over the resistor, which results in a model like

$$0 = (1)i + (-G)v + (0)$$
$$j = (0)i + (0)v + u + (0)$$
(5.2)

with G the conductance. Extension towards a nonlinear resistor is not difficult.

The model description for a transistor can be obtained using the Ebers-Moll replacement model as depicted in Fig. 5.3 where the currents can be expressed as

$$I_f = \left(V_{gs} - V_t\right)_+^2, I_r = \left(V_{gd} - V_t\right)_+^2$$

with V_t the threshold voltage where the diodes are ideal. When approximating the square functions using three segments, the $I_f - V_{gs}$ characteristic has a shape as like in Fig. 5.3. The overall static MOS transistor model has a shape like depicted in Fig. 5.4, which can be deduced from Fig. 5.3 by geometrical inspection. The PL model for the square function can be obtained via the Chebychev approximation as treated in section 5.1.1. With an error of $\delta = 0.16$ for x < 3.23 the model can be given by

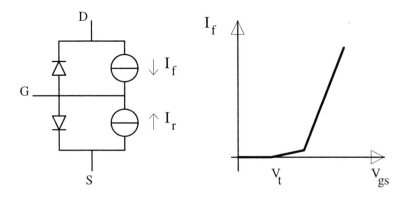

Figure 5.3. Ebers-Moll model (left) and the current characteristic (right)

$$0 = I_f + (0)V_{gs} + (-0.8, -2.26)u + (0)$$

$$j = \begin{pmatrix} -1 \\ -1 \end{pmatrix} V_{gs} + Iu + \begin{pmatrix} V_t + 0.97 \\ V_t + 2.10 \end{pmatrix} \qquad (5.3)$$

from which the complete function can be deduced with $I_{ds} = I_f - I_r$,

$$0 = I_{ds} + \begin{pmatrix} 0 & 0 & 0 \end{pmatrix} \begin{pmatrix} V_d \\ V_g \\ V_s \end{pmatrix} + \begin{pmatrix} -0.8 & -2.26 & 0.8 & 2.26 \end{pmatrix} u + (0)$$

$$j = \begin{pmatrix} 0 & -1 & 1 \\ 0 & -1 & 1 \\ 1 & -1 & 0 \\ 1 & -1 & 0 \end{pmatrix} \begin{pmatrix} V_d \\ V_g \\ V_s \end{pmatrix} + Iu + \begin{pmatrix} V_t + 0.97 \\ V_t + 2.10 \\ V_t + 0.97 \\ V_t + 2.10 \end{pmatrix} \qquad (5.4)$$

Inclusion of the channel length modulation effect as well as the body effect can relatively simple be achieved as will be demonstrated in section 5.3. The dynamics of the MOS transistor can be included by realizing a sub network consisting of the static MOS transistor and with the inclusion of capacitor models. The discontinuities in the slope of the characteristic cannot be noticed in the response during time analysis due to the averaging effect of the capacitance's.

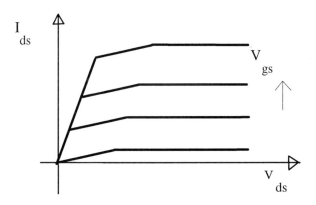

Figure 5.4. The static MOS transistor characteristic

5.1.4. Dynamic components

To cover dynamic components like capacitors the model description as proposed by van Bokhoven is extended by state variables z of the system (charge on capacitors and flux through coils) and their time derivatives $\dot{z}$,

$$0=Iy+Ax+Bu+Hz+J\,\dot{z}+f$$

$$0=Ky+Lx+Mz+N\,\dot{z}+r \tag{5.5}$$

$$j=Dy+Cx+Iu+g$$

and the complementary relation still holds. The first and last equation describe a PL function and therefore describe the static PL parts and the PL descriptions of the nonlinear relations of the dynamic parts. The middle equation is linear and describes the linear time dependent parts. For a nonlinear capacitor with the Q-V function as depicted in Fig. 5.5 the model can be given as

$$0=(1)v+(0)i+(3)u+(-2)z+(0)\dot{z}+(0)$$

$$0=(0)v+(1)i+(0)z+(-1)\dot{z}+(0) \tag{5.6}$$

$$j=(-1)v+(0)i+Iu+(2)$$

with the charge Q represented by z.
    In this way the components of the electrical level can be modeled with the model description of van Bokhoven.

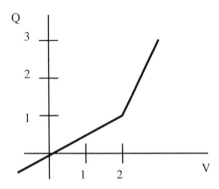

Figure 5.5. The Q-V relation of the nonlinear capacitor

5.1.5. Behavioral components

In behavioral modeling we try to abstract a network of electrical level models and group them into a single model. For example a filter is then not described in terms of resistors, capacitors and active components but as a differential equation.

A first order ordinary differential equation will look like

$$\dot{y} + a_0 y = b_1 \dot{x} + b_0 x$$

that can be rewritten as

$$y = b_1 x + z$$
$$\dot{z} = b_0 x - a_0 y$$

(5.7)

for which the model description will be

$$0 = (1)y + (-b_1)x + (0)u + (-1)z + (0)\dot{z} + (0)$$
$$0 = (a_0)y + (-b_0)x + (0)z + (-1)\dot{z} + (0)$$
$$j = (0)y + (0)x + u + (0)$$

(5.8)

In this way (nonlinear) differential equations of any order can be modeled.

As behavioral model, consider for instance also the behavioral model for the hysteresis curve as presented in chapter 2. In section 5.4 we will demonstrate the feasibility of behavioral or macro modeling for analyzing complex networks. Here we consider macro modeling as the concept of modeling the behavior of a single component using a network of several electrical and behavioral connected components.

5.1.6. Remarks

The possibilities to model a certain behavior on a abstract level demand some insight and feeling with the actual model as well as with piecewise linear modeling techniques. However, many signal operations can be modeled using PL techniques. There is only one limitation for the presented explicit model descriptions, the fact that the function must be continuous. For this reason, ideal switches, ideal logic components, et cetera cannot be modeled. Implicit model descriptions can handle an ideal step function, but the model is not simple analyzed by a PL simulator. Sometimes it is more convenient to use more component models to describe the complete behavior of the system. In that way a sub network is realized. For example, the extension of the static MOS transistor model towards a dynamic model results in a sub network for the dynamic transistor. In the previous chapter we have seen how such (sub) networks with PL component models can be analyzed.

5.2. Advanced modeling of combinatorial logic

Already in section 5.1 a piecewise linear model of a digital component was given. This model was derived using threshold logic. In this section we will present a method to model any combinatorial logic function, ranging from simple gates to high level building blocks like multiplexers, carry look-ahead circuits et cetera. The method automatically generates the model description out of the combinatorial function. To derive the model description, two concepts are combined, i.e. threshold logic and a method to solve a set of simultaneous inequalities. We will discuss them both. In the discussion below often the term separable is used, for which the definition is

Definition 5.1 *A combinatorial logic function with n input variables and one output variable y is called linear separable (l.s) if on the Boolean hypercube in R^n the 0 values of y can be separated from the 1 values of y using a single hyperplane.*

From this point of view an AND gate is linear separable but a XOR gate is not, as can be seen from Fig. 5.6.

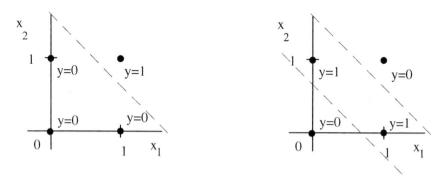

Figure 5.6. Separation of zero's and ones for an AND gate (left) and XOR gate (right)

Threshold logic is a very old technique to realize logic functions [75-76]. The basic element in this technique is the threshold gate, see Fig. 5.7. In this gate, the binary inputs i are weighted with factor a_i and summed. Using a lower and upper threshold, represented by l and u respectively, the binary output is determined,

$$y = 0 \qquad\qquad \text{if } \sum_{i=1}^{n} a_i x \le l$$

$$y = 1 \qquad\qquad \text{if } \sum_{i=1}^{n} a_i x \ge u \qquad\qquad (5.9)$$

$$y = \frac{1}{u-l}\left(\sum_{i=1}^{n} a_i x - l\right) \quad \text{elsewhere}$$

However, the realization of a logic function using a single threshold gate is only possible for separable functions only. Unfortunately, the majority of all logic functions is non-$l.s.$ It can be shown that of all 65,536 functions of four inputs and one output only 2.87% is $l.s.$ and this number decreases rapidly as the number of inputs increases. This means that for the majority of the logic function more than one hyperplane is necessary to separate the zero's from the one's on the Boolean n-cube.

Each nonlinear separable function can be expressed as a set of $l.s.$ parts, each part realized by a separate threshold gate. The outputs of all these parts will simply be added to result the output of the non $l.s.$ logic. Since the addition of several PL functions is again a PL function this division of non $l.s.$ parts into a summation of $l.s.$ parts results in a PL model of a non $l.s.$ logic function. Therefore, the problem is to obtain a PL model for a given $l.s.$ logic function. Consider a $l.s.$ logic function having n input variables. According to (5.9) the input word specifying a logic zero at the output has to satisfy

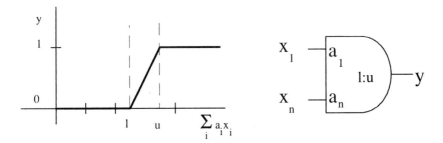

Figure 5.7. Threshold logic, the gate

$$\sum_{i=1}^{n} a_i x \leq l \qquad\qquad (5.10)$$

and a similar expression holds for the one's

$$\sum_{i=1}^{n} a_i x \geq u \qquad\qquad (5.11)$$

.

with $u > l$.

In this way, $2^n + 1$ inequalities are obtained from which the a_i's and the bounds have to be calculated. This can be done by applying the Tschernikow algorithm [39]. With this algorithm, a set of homogeneous inequalities given by

$$Ax \leq 0, \quad x \geq 0, \quad A \in R^{mxn} \qquad\qquad (5.12)$$

can be solved, where the solution is expressed as

$$x = \sum_{i=1}^{k} p_i v_i \qquad \forall_{i \in \{1..k\}} p_i \geq 0 \qquad\qquad (5.13)$$

The solution space is described as a positive linear combination of a collection of vectors v_i, representing the corner points of the space (see also chapter 3 and 6). With this method, an algorithm can be developed which automatically generates a PL model of any logic function [77].

As example, consider the function of Fig. 5.8 that is a four stage carry look-ahead. The logic function has nine inputs and five outputs [78]. The inputs P_i, G_i and C_n are outputs of the four ALU's or adders (carry Propagate and carry Generate and first carry). The outputs C_{n+x}, C_{n+y} and C_{n+z} are the anticipated carry signals. The outputs P and G can be used as an input for another carry look-ahead unit. Equation (5.12) has been solved resulting in the values for u, l and a_l as depicted also in the figure.

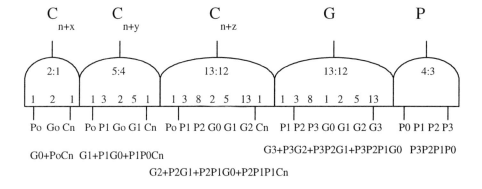

Figure 5.8. A four stage carry look-ahead circuit

The PL model for each part is now simply generated. For example, for C_{n+x} the hyperplanes according to (5.10), (5.11) and Fig. 5.7 are

$$P_o + 2G_o + C_n = 2$$
$$P_o + 2G_o + C_n = 1$$

The slope in the transit region is defined as $1/(u-l)=b=1$, hence the B matrix is given as $B=(-b,b)$, and if we start the model with the output equals zero then the A matrix has zero entries as well as f. Therefore the model is given as

$$0 = C_{n+x} + \begin{pmatrix} 0 & 0 & 0 \end{pmatrix} \begin{pmatrix} P_0 \\ G_0 \\ C_n \end{pmatrix} + \begin{pmatrix} -1 & 1 \end{pmatrix} u + (0)$$

$$j = \begin{pmatrix} -1 & -2 & -1 \\ -1 & -2 & -1 \end{pmatrix} \begin{pmatrix} P_0 \\ G_0 \\ C_n \end{pmatrix} + Iu + \begin{pmatrix} 2 \\ 1 \end{pmatrix}$$

5.3.   PL model for an nMOS transistor

In section 5.1.3 we already discussed a simple model for the MOS transistor. However, no channel length modulation effect, the dependency of the threshold voltage for the bulk voltage etc., were included. We will now discuss a more advanced model.

The MOS transistor, as depicted in Fig. 5.9 can be seen as a 4 terminal device (D = drain, G = gate, S = source, B = bulk). In the PL model, the bulk terminal will be used as the reference voltage. The voltage V_g represents the gate voltage minus the bulk voltage and so on.

For the basic behavior of the MOST, that is without channel length modulation and body effect, we will use the Sah model,

$$I_d = -I_s = \frac{K}{2} \cdot \left(f\left(V_g - V_s - V_T \right) - f\left(V_g - V_d - V_T \right) \right)$$

$$\begin{cases} x < 0 \rightarrow f(x) = 0 \\ x > 0 \rightarrow f(x) = x^2 \end{cases} \tag{5.14}$$

$$V_T = \text{threshold voltage,} \quad K = \frac{\mu \cdot C_{ox} \cdot W}{L}$$

When we write this model explicitly for the 4 operation regions, we get the following well-known equations,

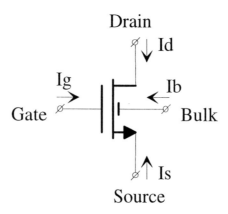

Figure 5.9. nMOST

$$\begin{cases} I_{ds} = 0 & \text{off: } (V_{gs} < V_T) \wedge (V_{gd} < V_T) \\ I_{ds} = \dfrac{K}{2} \cdot (V_{gs} - V_T)^2 & \text{forward saturation: } (V_{gs} > V_T) \wedge (V_{gd} < V_T) \\ I_{ds} = K \cdot \left\{ (V_{gs} - V_T) \cdot V_{ds} - \tfrac{1}{2} \cdot V_{ds}^2 \right\} & \text{linear or triode: } (V_{gs} > V_T) \wedge (V_{gd} > V_T) \\ I_{ds} = -\dfrac{K}{2} \cdot (V_{gd} - V_T)^2 & \text{reverse saturation: } (V_{gs} < V_T) \wedge (V_{gd} > V_T) \end{cases}$$

$$(5.15)$$

To make a PL model of the Sah model, we only have to approximate the nonlinear function f in (5.14) by a PL function. In order to have an accurate model, we make the following demands:

• for $x \le 0 \to PL(x) = 0$

• for $0.1 \le x \le 4 \to \dfrac{\left| PL(x) - x^2 \right|}{x^2} < 0.1 \cdot$

We can realize such a PL function with 5 hyperplanes, resulting in 6 polytopes. The breaking points in the function with corresponding approximated function values are at (0, 0), (0.1, 0.009), (0.2487, 0.0557), (0.6185, 0.3443), (1.5383, 2.1297) and the end point is (3.8256, 13.172). This PL function is plotted together with x^2 in Fig. 5.10.

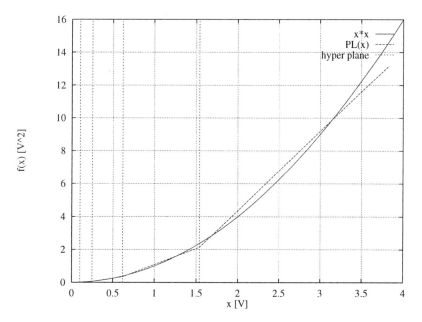

Figure 5.10. PL square function

The ten hyperplanes of the PL mapping of the nMOST model are equal to,

$$V_g - V_s = V_T - a_i$$
$$V_g - V_d = V_T - a_i$$
$$a_1 = 0, \ a_2 = 0.1, \ a_3 = 0.2487, \ a_4 = 0.6182, \ a_5 = 1.5383, \ i=1\cdots5$$

$$(5.16)$$

which are obtained using the Chebychev approximation (see section 5.1.1). The update values are related to the differences in slope of the mappings on both sides of the break-points. The PL-model of this basic nMOST is equal to,

$$0 = I \begin{bmatrix} I_d \\ I_g \\ I_s \end{bmatrix} + \begin{bmatrix} 0 & 0 & 0 \\ 0 & 0 & 0 \\ 0 & 0 & 0 \end{bmatrix} \begin{bmatrix} V_d \\ V_g \\ V_s \end{bmatrix} + \begin{bmatrix} -c_1 & -c_2 & \cdots & -c_5 & c_1 & \cdots & c_5 \\ 0 & 0 & \cdots & 0 & 0 & \cdots & 0 \\ c_1 & c_2 & \cdots & c_5 & -c_1 & \cdots & -c_5 \end{bmatrix} u + 0$$

$$j = \begin{bmatrix} 0 & -b & b \\ 0 & -b & b \\ \cdots & \cdots & \cdots \\ 0 & -b & b \\ \hline b & -b & 0 \\ \cdots & \cdots & \cdots \\ b & -b & 0 \end{bmatrix} \cdot \begin{bmatrix} V_d \\ V_g \\ V_s \end{bmatrix} + Iu + \begin{bmatrix} hyp_1 \\ hyp_2 \\ \cdots \\ hyp_5 \\ hyp_1 \\ \cdots \\ hyp_5 \end{bmatrix}$$

$$(5.17)$$

with

$$u \geq 0, \quad j \geq 0, u^T j = 0$$
$$b = \frac{K \cdot W}{2 \cdot L}, \quad A = \frac{b \cdot \gamma}{2\sqrt{\varphi}}$$
$$c_1 = 0.09, \ c_2 = 0.2238, \ c_3 = 0.4666, \ c_4 = 1.1605, \ c_5 = 2.8863$$
$$hyp_1 = b \cdot V_T, \ hyp_2 = b \cdot (V_T + 0.1), \ hyp_3 = b \cdot (V_T + 0.2487)$$
$$hyp_4 = b \cdot (V_T + 0.6185), \ hyp_5 = b \cdot (V_T + 1.5383)$$

and with K, γ, ϕ, V_T process dependent.

The Early effect is usually included by multiplying the drain-source current by a drain-source voltage dependent factor:

$$I_{ds} = I_{ds,SAH} \cdot \left(1 + \left|\lambda \cdot V_{ds}\right|\right)$$ (5.18)

In PL models, a multiplication of two variables has to be approximated by means of several polytopes. We therefore model the Early effect in a somewhat different way (which is in first order equal to the usual way),

$$I_d = -I_s = \frac{K}{2} \cdot \left(f\left(V_{gs} - V_T + \tfrac{1}{2} \cdot \lambda \cdot V_{ds} \cdot \left(V_{gs} - V_T\right)\right) - \right.$$
$$\left. f\left(V_{gd} - V_T - \tfrac{1}{2} \cdot \lambda \cdot V_{ds} \cdot \left(V_{gd} - V_T\right)\right)\right)$$ (5.19)

with $f(x)$ as in (5.14).

When we take $V_{gs} - V_T$ in the Early voltage part equal to a reference voltage (equal to the average value in the particular polytope), we do not have to increase the number of polytopes, to include the Early effect. The only parameters that changes are the hyperplanes (the updates can remain the same),

$$V_g - V_s \cdot \left(1 + \lambda \cdot \left(\frac{a_{i+1} - a_i}{4}\right)\right) + V_d \cdot \lambda \cdot \left(\frac{a_{i+1} - a_i}{4}\right) = V_T - a_i$$

$$V_g - V_d \cdot \left(1 + \lambda \cdot \left(\frac{a_{i+1} - a_i}{4}\right)\right) + V_s \cdot \lambda \cdot \left(\frac{a_{i+1} - a_i}{4}\right) = V_T - a_i$$ (5.20)

$$a_1 = 0, a_2 = 0.1, a_3 = 0.2487, a_4 = 0.6182$$
$$a_5 = 1.5383, a_5 = 3.8256, i = 1 \cdots 5$$

The body effect can in first order be modeled as a linear relation between the channel-bulk voltage and the threshold voltage. This effect also only requires a modification in the equations for the hyperplanes

$$V_g - V_s \cdot \left(A + \lambda \cdot \left(\frac{a_{i+1} - a_i}{4}\right)\right) + V_d \cdot \lambda \cdot \left(\frac{a_{i+1} - a_i}{4}\right) = V_T - a_i$$

$$V_g - V_d \cdot \left(A + \lambda \cdot \left(\frac{a_{i+1} - a_i}{4}\right)\right) + V_s \cdot \lambda \cdot \left(\frac{a_{i+1} - a_i}{4}\right) = V_T - a_i$$ (5.21)

$$a_1 = 0, a_2 = 0.1, a_3 = 0.2487, a_4 = 0.6182$$
$$a_5 = 1.5383, a_5 = 3.8256, i = 1 \cdots 5$$

The factor A in (5.21) is the body-effect constant, which is typically in the order of 1.1.

When the body-effect as well as the Early-effect are included in the basic PL model (5.17), the model becomes equal to

$$
0 = I\begin{bmatrix} I_d \\ I_g \\ I_s \end{bmatrix} + \begin{bmatrix} -0 & 0 & +0 \\ 0 & 0 & 0 \\ +0 & 0 & -0 \end{bmatrix}\begin{bmatrix} V_d \\ V_g \\ V_s \end{bmatrix} + \begin{bmatrix} -c_1 & -c_2 & \cdots & -c_5 & c_1 & \cdots & c_5 \\ 0 & 0 & \cdots & 0 & 0 & \cdots & 0 \\ c_1 & c_2 & \cdots & c_5 & -c_1 & \cdots & -c_5 \end{bmatrix} u + 0
$$

$$
j = \begin{bmatrix} -a_1 & -b & A+a_1+b \\ -a_2 & -b & A+a_2+b \\ \cdots & \cdots & \cdots \\ -a_5 & -b & A+a_5+b \\ \hline A+a_1+b & -b & -a_1 \\ \cdots & \cdots & \cdots \\ A+a_5+b & -b & a_5 \end{bmatrix}\begin{bmatrix} V_d \\ V_g \\ V_s \end{bmatrix} + Iu + \begin{bmatrix} hyp_1 \\ hyp_2 \\ \cdots \\ hyp_5 \\ hyp_1 \\ \cdots \\ hyp_5 \end{bmatrix} \tag{5.22}
$$

with

$$u \geq 0,\ j \geq 0,\ u^T j = 0$$

$$b = \frac{K \cdot W}{2 \cdot L},\quad A = \frac{b \cdot \gamma}{2\sqrt{\varphi}}$$

$c_1 = 0.09,\ c_2 = 0.2238,\ c_3 = 0.4666,\ c_4 = 1.1605,\ c_5 = 2.8863$

$a_1 = b \cdot \lambda \cdot 0.025,\ a_2 = b \cdot \lambda \cdot 0.0872,\ a_3 = b \cdot \lambda \cdot 0.2168,\ a_4 = b \cdot \lambda \cdot 0.5392$

$a_5 = b \cdot \lambda \cdot 1.341,\ hyp_1 = b \cdot V_T,\ hyp_2 = b \cdot (V_T + 0.1),\ hyp_3 = b \cdot (V_T + 0.2487)$

$hyp_4 = b \cdot (V_T + 0.6185),\ hyp_5 = b \cdot (V_T + 1.5383)$

Fig. 5.11 depicts the large signal behavior of an nMOST with W = L = 2.4 μm. The source is connected to the bulk, and the process parameters are: $\lambda = 1.5151e\text{-}7$, L = 0.063 (for L = 2.4 μm), $\varphi = 0.7$, $\gamma = 0.3$ and K = 57e-6.

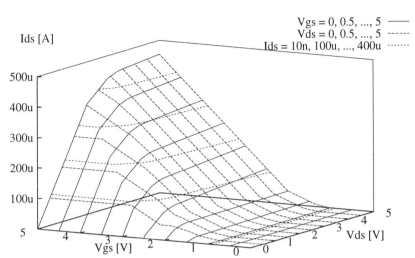

MIETEC 2.4u CMOS, nMOST W/L = 2.4/2.4

Figure 5.11. PL model of a minimal size nMOST.

Although the analytical MOS models are more accurate than this PL model, the latter is in many applications accurate enough. The advantage is that transistor networks where the components are modeled with such PL models, can be faster analyzed (in terms of CPU time). Inclusion of the dynamics is relative simple by connecting capacitance's to the static model. For many MLMS simulations the in this way obtained PL model will do.

5.4.   A PL macro model for a CSA circuit

Here the development of an accurate PL model description for an Charge Sensitive Amplifier (CSA) will be discussed as an example of macro modeling. The function of CSA is to integrate all charge that is presented at the input on a capacitor. The output voltage of the CSA has to be equal to the reference voltage minus the voltage over the capacitor. In all pulse electronic systems, a CSA is realized by connecting a capacitor in the feedback loop of an inverting amplifier as depicted in Fig. 5.12.

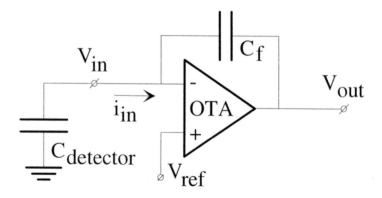

Figure 5.12. Principle schematic of a CSA

This basic schematic is also used to generate the macro model. Besides the charge to voltage function of the CSA, several of the main non idealities are modeled as well. The most important negative second order effect is the noise contribution of the CSA. As stated in [79], the noise contribution of the CSA is dependent on the following parameters: input voltage noise of the OTA, the g_m (transconductance) of the OTA, C_f and all parasitic capacitance's connected to the input (including the detector capacitance). Other non ideal effects are a limited output voltage, a non-zero rising time of the output signal, and a drop in the output voltage. The parameters that are responsible for these effects can be specified in the model of the CSA. The macro model is a small network of PL components, which in SPICE-like syntax can be given as

CSA

OTA ref in_ota out

Cf out in

Rf out in

Vnoise in in_ota

Cp in ground

This model of a CSA is in a hierarchical simulator like PLANET seen as a sub network of the complete circuit where the CSA is just one element of the circuit. The component Vnoise is a noise voltage source. The component OTA is a macro model of an operational transconductance amplifier. The relation between the output current and the input voltages is given by

$$i_{out} = -g_m \left(v_+ - v_- \right) - g_{ds} v_o$$

However, the OTA will be saturated for currents larger than I_0, which can be modeled using two hyperplanes at

$$v_+ - v_- = I_0 / g_m \text{ and } v_+ - v_- = -I_0 / g_m$$

Also the output voltage is limited, resulting in two other planes, namely $v_0 = V_{max}$ and $v_0 = -V_{min}$. The complete PL model description for this OTA can then be given as

$$0 = I \begin{pmatrix} i_0 \\ i_+ \\ i_- \end{pmatrix} + \begin{pmatrix} -g_{ds} & -g_m & g_m \\ 0 & 0 & 0 \\ 0 & 0 & 0 \end{pmatrix} \begin{pmatrix} v_0 \\ v_+ \\ v_- \end{pmatrix} + \begin{pmatrix} g_m & -g_m & 0 & 0 \\ 0 & 0 & 0 & 0 \\ 0 & 0 & 0 & 0 \end{pmatrix} u + \begin{pmatrix} 0 \\ 0 \\ 0 \end{pmatrix}$$

$$j = \begin{pmatrix} 0 & -1 & 1 \\ 0 & 1 & -1 \\ -1 & 0 & 0 \\ 1 & 0 & 0 \end{pmatrix} \begin{pmatrix} v_0 \\ v_+ \\ v_- \end{pmatrix} + Iu + \begin{pmatrix} I_0 / g_m \\ I_0 / g_m \\ V_{max} \\ V_{min} \end{pmatrix}$$

$$(5.23)$$

The simulated output voltage of the CSA when used with a sensor capacitance of 100 pF, an input charge of -1.0 fC (at t = 1 µs), a DC-gain of 1000 and all other parameters equal to their default values is depicted in Fig. 5.13.

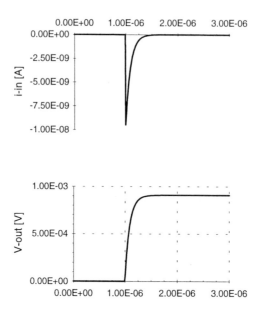

Figure 5.13. Simulation results of the CSA.

As can be seen in the simulation results, the charge coming from the detector is not transferred immediately to the integrating capacitor Cf, but it takes about 0.4 µs before most of the charge is transferred. Furthermore, it can be seen that the output voltage of the CSA does not reach the ideal value of 1 mV, but only becomes 0.91 mV. The reason for the latter effect is the limited DC-gain, combined with the ratio between the integrating capacitor and the detector capacitance. This effect is therefore not a simulation error, but an effect that will occur in "real-life" circuits as well. It is very important that these effects are visible in the simulation results, since these effects make that the circuit to be designed is not operating as it should be.

5.5.   Complex network examples

In this section we will show the usefulness of MLMS simulations with a piecewise linear simulator. The used simulator is PLANET [59] running on a HP700 workstation.

5.5.1. A phase-locked loop circuit

An example of a MLMS network is a phase-locked loop (PLL). A PLL is always a difficult network to analyze due to the double nested feedback loop, one is the PLL main loop and one is in the VCO and the fact that the two related time constant are far away of each other. Although the PLL itself is highly nonlinear, the building blocks are easily obtained. In Fig. 5.14 the PLL contains three sub networks:

- *A phase detector (PD)*. The PD is modeled as an exclusive-or function using the techniques outlined in section 5.2. It inputs and outputs are modeled on the binary level.
- *A loop filter (LF)*. The first order loop filter is modeled as a ordinary differential equation like is done in section 5.1.5. Therefore this component is modeled at behavioral level.
- *A voltage-controlled oscillator (VCO)*. The VCO is a small network consisting of components, modeled at the electrical level.

Looking at the components of the PLL, it can be concluded that indeed the PLL is a mixed level mixed signal network. If the structure of the circuit is depicted like in Fig. 5.15, then also hierarchical simulation can be performed.

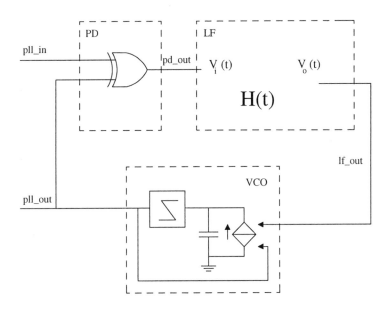

Figure 5.14. The PLL as a MLMS network

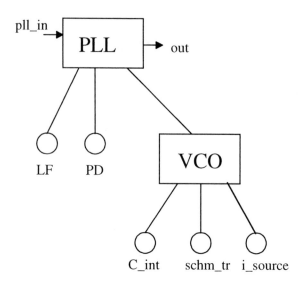

Figure 5.15. Hierarchy of the PLL used within PLANET. The i_source represents the controlled current source, schm_tr the hysteresis curve and C_int the capacitor as depicted in Fig. 5.14. Hierarchy is obtained by separation of the blocks LF, PD and VCO with respect to their topological equations.

The results of the transient simulations are given in Fig. 5.16 and illustrate the pull-in behavior of the PLL. The input of the PLL is a 1500 Hz square wave. Initially, the VCO runs at a frequency of 1000 Hz but this is gradually changed to a frequency equal to the input frequency by the DC component of the signal at the output of the loop filter. The CPU time was 13 seconds for 2500 time points and the size of the data structure of the tree is less than 2000 floating point entries. This example also demonstrates the good global convergence properties of the simulator because the given circuit is computationally difficult and many network simulators will break down on it.

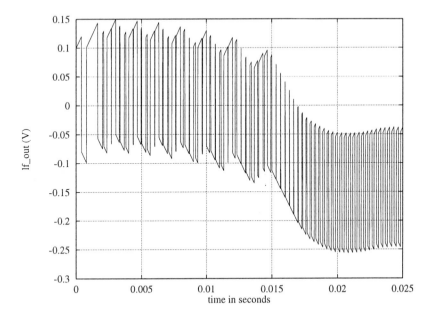

Figure 5.16. Pull-in behavior of the PLL

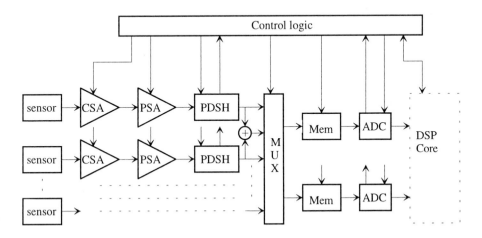

Figure 5.17. Particle detector read-out system

5.5.2. Particle detection system

As last example we will demonstrate the usefulness of the PL concept in MLMS application by means of the analysis of a particle detection system.

A typical particle detector read-out system is depicted in Fig. 5.17. Considering the complexity it is time consuming and not desirable to simulate the entire system at the transistor level. It is therefore required to have a suitable set of macro or behavioral models with which the network can be analyzed.

The following blocks are needed:

- *Control Logic.* The digital blocks in the control logic can be built with standard available gates like AND, NAND, OR, NOR and INV.
- *Analog / Digital Converter (ADC)* For our purpose, we have modeled an ADC that does not require any clock signals and which has a conversion time of 0 seconds. On the basis of this converter, other ADC types can be modeled by adding small circuits like for instance a sample switch and a capacitor in order to get a clocked ADC.
- *Pulse Shaping Amplifier (PSA)* There are several different types of pulse shaping amplifiers of which the semi-Gaussian is used most often. This type of PSA is a cascade of a first order differentiator and a n-th order integrator. The time constants of the integrator and the differentiator are equal to the peaking time divided by the order (n) of the PSA. A block diagram of the semi-Gaussian PSA is depicted in Fig. 5.18. Besides this basic functionality, several other parameters can be specified. The minimal and maximal values of the output voltage can be specified and furthermore, the gain and an optional output impedance. The simulation results obtained by PLANET for a PSA with order n = {2, 4, 6, 20} and all other parameters equal to their default values are depicted in Fig. 5.19. The input signal is a step voltage of 1 V at t = 0 μs.

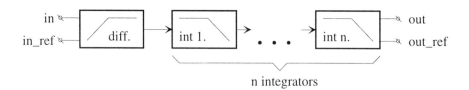

Figure 5.18. Semi Gaussian Pulse Shaping Amplifier

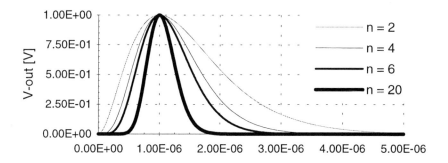

Figure 5.19. Simulated PSA output

- *Peak Detect Sample & Hold (PDSH)* A peak detect sample and hold circuit is used to track a signal when the signal is rising and holds the peak value when the signal is falling. Several different principles are possible to realize such a behavior. In the system, a PDSH is used that is based on the circuit [80]. This kind of PDSH only requires one control input signal (the reset signal). The status of the PDSH (the PDSH is either in the 'track-mode' or in the 'hold-mode') can easily be detected by comparing the input and the output voltage of the PDSH. This is not implemented in the used model, but only requires one additional comparator. The PDSH with as parameters a dynamic range of 10 mV/μs, a slew rate of 1 V/ns, and a gain-bandwidth of 10 MHz is simulated with PLANET. The input of the PDSH is a 4-th order semi-Gaussian pulse, with a peaking time of 1 μs, and a peak value of 1 V. A reset pulse was presented at t = 0 μs and at t = 6 μs. The simulation result is depicted in Fig. 5.20. As depicted in Fig. 5.20, the output voltage of the PDSH does not reach the peak value of the input voltage, due to the limited gain bandwidth of the simulated PDSH. When the PDSH is in the 'hold-mode' (2.2 μs < t < 6.0 μs), the output voltage slowly decreases due to the non-zero droop rate. These effects will also occur in "real-life" circuits. It is therefore important to analyze these effects with simulations, in order to prevent unpleasant surprises after a chip has been produced.

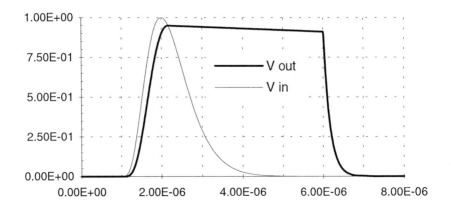

Figure 5.20. Simulation result of the PDSH model

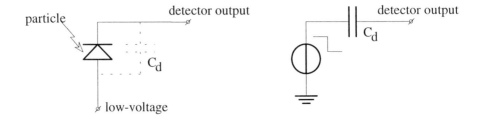

Figure 5.21. Modeled particle detector behavior

- *Particle detector.* The particle detectors produce a short current pulse each time the detector is hit by a certain particle. The amount of charge coming out of the detector is proportional to the energy of that particle. Since the detector output is connected to the CSA input (which is a virtual ground), the behavior of the sensor can be modeled by a capacitor, connected to a step voltage as depicted in the right-hand side of Fig. 5.21. Each time that a particle enters the detector, the voltage source makes a negative step, proportional to the energy of the particle. Since the model is very small, it is not defined as a model in the PLANET library, but we simply connect a capacitor to a step voltage source.

- *Charge Sensitive Amplifier (CSA).* This component is already discussed in the previous section.

The simulated pulse electronic system has two input channels. When a particle is detected by one of the channels, the output of that channel is connected to the ADC by means of the multiplexer (see also [81]). The ADC converts the analog signal to a 5 bits digital word, proportional to the energy of the particle. Besides the output of the ADC, there are two more digital output signals: 'detect' and 'select'. The signal 'detect' is high between the moment that a particle is detected and the moment that a reset pulse is provided to the system. When no particle is detected, the signal is low. The signal 'select' is high when a particle is detected by channel 1 and low when the particle is detected by channel 0. Two particles are provided as the input signal: the first hit detector 0 at $t = 2$ µs, while the second particle hits detector 1 at t $= 9$ µs. A block diagram of the system is depicted in Fig. 5.22, together with the most important signal waveforms obtained after simulation. The entire simulation took 90 CPU seconds and predicts very well the operating behavior of the entire system.

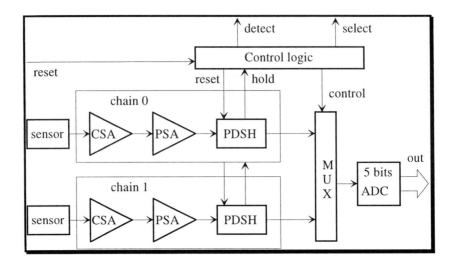

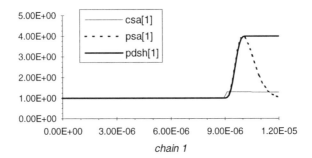

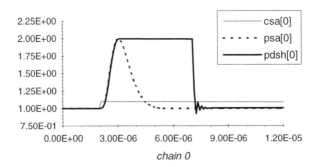

Figure 5.22. Simulated results. The system configuration and the signals in chain 0 and 1.

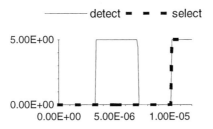

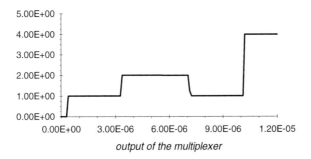

output of the multiplexer

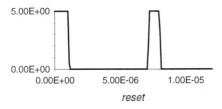

reset

Figure 5.22. Simulated results. The detection, select and reset signals and the output of the multiplexer.

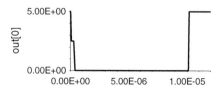

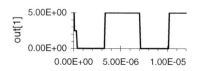

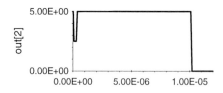

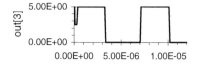

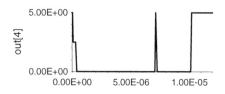

Figure 5.22. Simulated results. The binary output signals of the 5 bits ADC.

5.6. Automatic model generation for canonical model descriptions

In 1986 Chua presented a method that allowed to generate a piecewise linear model from a given set of data points [74]. The method is only valid for the canonical representation

$$y = a + bx + \sum_{i=1}^{\sigma} c_i |x - \beta_i| \qquad (5.24)$$

where we assume here a one-dimensional behavior. In the paper it is outlined how the method can be extended towards higher orders.

Given a set of N data points we want to find the set of parameters such that the approximation error

$$E(z_1, z_2) = \sum_{k=1}^{N} \left[w^k \left(a + bx^k + \sum_{i=1}^{\sigma} c_i |x^k - \beta_i| - y^k \right)^2 \right] \qquad (5.25)$$

is minimized, where

$$z_1 = \begin{pmatrix} a & b & c_1 & c_2 & \cdots & c_\sigma \end{pmatrix}^T \qquad (5.26)$$

$$z_2 = \begin{pmatrix} \beta_1 & \beta_2 & \cdots & \beta_\sigma \end{pmatrix}^T \qquad (5.27)$$

and w^k represents the weighting factor for data point k.

Assume that the location of each breakpoint is fixed at $z_2 = \hat{z}_2$ then obviously the approximation error (5.25) is a quadratic function of z_1 and the minimum is found by solving the linear equation

$$\frac{dE(z_1, \hat{z}_2)}{dz_1} = 2AWr = 0 \Rightarrow AW^{1/2}W^{1/2}A^T z_1 - AWy = 0 \qquad (5.28)$$

where

$$A = \begin{pmatrix} 1 & 1 & \cdots & 1 \\ x^1 & x^2 & \cdots & x^N \\ u_1^1 & u_1^2 & \cdots & u_1^N \\ \vdots & \vdots & \cdots & \vdots \\ u_\sigma^1 & u_\sigma^2 & & u_\sigma^N \end{pmatrix}, \qquad r = \begin{pmatrix} r^1 \\ r^2 \\ \vdots \\ r^N \end{pmatrix} \tag{5.29}$$

and

$$W = diag(w^1, w^2, \ldots, w^N)$$

$$r^k = a + bx^k + \sum_{i=1}^{\sigma} c_i |x^k - \beta_i| - y^k \tag{5.30}$$

$$u_i^k = |x^k - \beta_i|$$

The solution $z_1 = z_1^*$ is optimal for the fixed partition $z_2 = \hat{z}_2$. Different choices of the partitioning will give different optimal parameter sets. The goal is now to choose the optimal partitioning or

$$E\left(z_1^*(\hat{z}_2), \hat{z}_2\right) = \min\left\{ E\left(z_1^*(z_2), z_2\right) \middle| z_2 \in R^\sigma \right\} \tag{5.31}$$

Define

$$g = \frac{\partial E(z_1, z_2)}{\partial z_2}, \qquad Y = \frac{\partial g}{\partial z_2} \tag{5.32}$$

then we can perform a line search along the direction $s = -Y^{-1}g$ which is guaranteed to be in the descent direction since Y is positive definite. The result is the following iteration procedure that starts from the initial partition $z_2 = z_2^0$. We find α^k such that

$$E\left(z_1^*(z_2^k + \alpha^k s^k), z_2^k + \alpha^k s^k\right) = \min\left\{ E\left(z_1^*(z_2^k + \alpha z_2^k), z_2^k + \alpha s^k\right) \middle| \alpha \geq 0 \right\} \tag{5.33}$$

and we update $z_2^{k+1} = z_2^k + \alpha^k s^k$.

This procedure reduces the error E for each iteration and hence must approach a local minimum. However, there may exist more than one local minimum and therefore it is not guaranteed that the optimal solution will be found.

The above presented method can be extended towards higher dimensional problems. The procedure is robust and will always come up with a model. The problem with the method is that the on forehand it can not be said how accurate the model will be. There is no error control in it and it is not possible to define at the start of the procedure what the overall error should be. Further, because the underlying model description is limited towards the class of function that it can handle, only rather simple characteristics can be modeled automatically.

5.7. Automatic model generation for implicit model descriptions

One method to overcome the drawbacks of the previous method is the one proposed by Veselinovic [90]. The method relies on model description *Bokh2*, hence can handle a broader class of functions. Further the method makes use of error propagation and allows the user to specify an overall accuracy for a certain interval for the output and input variables.

5.7.1. The model generator

A nonlinear device is described as a set of nonlinear functions which maps input variables onto the output variables,

$$y_i = f_i(x_1, x_2, \cdots, x_m)$$
(5.34)

The functions f_i are required to be in an explicit form and currently they may be composed of following operators $\{+, -, *, /, \wedge, log, exp\}$. Each of the functions f_i is approximated according to the proposed method and in the rest of this section we will be talking about the overall (input) function having in mind one of the functions f_i.

The aim is to find a PL approximation which in the worst case deviates from the overall function less than what a given relative error (e_{max}) allows. All the approximations are valid within a certain region, therefore the domains for all the input variables should be known.

We begin with a given nonlinear multivariable function, a given maximal allowed relative error and a given set of intervals for input variables. The function is then being decomposed into factors, each of that is either a sum of two factors or a single-variable function, with the similar restriction for the choice of operators as above. The decomposition follows the rules of the operator priority in a recursive descent syntax analysis and parses the function into factors. The factor's intervals are calculated from the initial intervals (domain) for the overall function's variables. The relative error for each factor is calculated starting with the given relative error of the whole function. The simplification of the function is done by evaluating the contribution of each factor and its assigned relative error. Depending on the size of the allowed error, some factors are negligible with respect to the overall function. In order to have an efficient PL model, the number of factors must be kept minimal for a given relative error.

To start with, the following has to be defined: the *function*, for example

$$y = a \cdot x_1^2 \cdot \left(b \cdot x_2 + c \cdot x_3\right) \cdot e^{d \cdot x_4} \tag{5.35}$$

the *interval* of the values for all used variables

$$28 \le x_1 \le 43$$
$$39 \le x_2 \le 52$$
$$8 \le x_3 \le 13 \tag{5.36}$$
$$2 \le x_4 \le 2.1$$

the *parameter values* (for example)

$$a = 1.3$$
$$b = 2$$
$$c = 0.53 \tag{5.37}$$
$$d = 3.14$$

and the *overall relative error*, e_{max}=1%.

The goal of decomposition is a set of factors for which the PL approximation is a straightforward task. A parser based on left-to-right *recursive descent* syntax analysis, respecting the arithmetic operator priority, is used to decompose the expression (5.35). The decomposition of (5.35) into factors yields,

$$y_1 = \ln x_1; \; y_2 = y_1 \cdot 2; \; y_3 = e^{y_2}; \; y_4 = a \cdot y_3; \; y_5 = b \cdot x_2;$$
$$y_6 = c \cdot x_3; \; y_7 = y_6 + y_5; \; y_8 = \ln y_4; \; y_9 = \ln y_7;$$
$$y_{10} = y_8 + y_9; \; y_{11} = e^{y_{10}}; \; y_{12} = d \cdot x_4; \; y_{13} = e^{y_{12}}; \tag{5.38}$$
$$y_{14} = \ln y_{11}; \; y_{15} = \ln y_{13}; \; y_{16} = y_{14} + y_{15}; \; y_{17} = e^{y_{16}}$$

and Fig. 5.23 shows a graph for this decomposition. A simple multiplication can be, in principle, transformed as

$$x_1 \cdot x_2 = e^{\ln x_1 + \ln x_2} \tag{5.39}$$

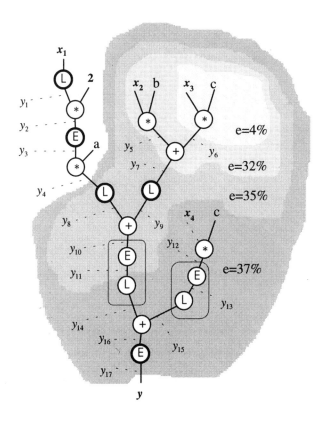

Figure 5.23. Example of pruning for function (5.35); nodes L and E are logarithm and exponent operators

and then parsed accordingly. This has been done for factors (y_4, y_7) and (y_{11}, y_{13}). Divisions and powers are treated in a similar form. The constant values and variable's domain is of no concern for the decomposition. It is, however, crucial for the assessment of the allowed error attributed to a factor. Variables' domain in case of negative values is a problem for transformation of (5.35), though. The solution is presented later on.

Intervals are calculated top-down (Fig. 5.23), beginning with leaf nodes and using the intervals of incoming nodes.

When a node has one incoming branch, its function is performed upon the interval that comes by the incoming branch. A plus node has an interval that is the sum of the incoming intervals, a minus node e.g. x_1-x_2 has an interval like $(x_{1min}$-x_{2max}, x_{1max}-$x_{2min})$. Plus and minus nodes produced by decomposition of *, /,^ are simple minimum and maximum function of incoming intervals.

Relative errors are calculated bottom-up (Fig. 5.23), starting with the relative error for the overall function and calculating errors allowed for each factor. All the

170 PIECEWISE LINEAR MODELING

factors will get the largest possible relative errors, that still ensure that in the worst case the given relative error of the overall function will not be violated. A general case for error calculation for an incoming node is

$$y = f(x); \quad e_x = \left|\frac{1}{x}\right| \cdot \left| f^{-1}\left(\left(1 \pm e_y\right) \cdot f(x)\right) - x \right| \tag{5.40}$$

Relative error propagated to the incoming node of an E node ($y = exp(x)$) and an L node ($y = log(x)$) are given by

$$e_x = \frac{\ln\left(1 + e_y\right)}{\max\left(\left|x_{min}\right|, \left|x_{max}\right|\right)} \tag{5.41}$$

and

$$e_x = \left|x_m^{\pm e_y} - 1\right|, \quad x_m = \begin{cases} x_{min}; & x_{min} > 1 \\ x_{max}; & x_{max} < 1 \end{cases} \tag{5.42}$$

respectively. Relative errors propagated through a plus or minus node y to its respective incoming nodes x_1 and x_2 are given by

$$e_{x_1} = \left|\frac{e_y}{2} \cdot \left(1 + \text{sgn}(y) \cdot \frac{\min\left(\left|x_{2min}\right|, \left|x_{2max}\right|\right)}{\max\left(\left|x_{1min}\right|, \left|x_{1max}\right|\right)}\right)\right| \tag{5.43}$$

$$e_{x_2} = \left|\frac{e_y}{2} \cdot \left(1 + \text{sgn}(y)\frac{\min\left(\left|x_{1min}\right|, \left|x_{1max}\right|\right)}{\max\left(\left|x_{2min}\right|, \left|x_{2max}\right|\right)}\right)\right| \tag{5.44}$$

The sign of y is considered over its whole domain, thus the problem occurs when the domain of y includes zero.

The problem of decomposition is that the logarithm function is undefined for arguments that are smaller than or equal to zero. Knowing that, (5.39) is valid only for positive values x_1 and x_2. Figure 5.24 shows a decomposition and transformation of a product of two variables, valid for negative as well as for positive values. Similar decomposition and transformation are required for x_1/x_2 as well as for $x_1{}^{\wedge}x_2$.

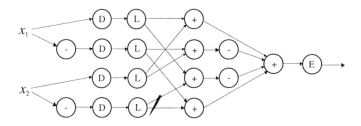

Figure 5.24. Decomposition example for multiplication in 4 quadrants; symbol (D) stands for an operator that allows only positive values (ideal diode)

The problem with the interval calculation is that the intervals for each node's domain are calculated "locally" as if the node's variables were independent. For example, had (5.35) the same form but only 3 variables, (where x_4 is substituted by x_3), the proposed method for interval calculation would yield in an unnecessarily larger interval for y_{16} than it actually ought to be. This might lead to the problem discussed later on as well as to a non-minimal number of linear segments in the final approximation. A possible solution is to numerically explore the domains at the moment of function decomposition, for all the (sub) branches of the graph (Fig 5.23). That would guarantee the minimal interval and the optimal PL approximation.

The problem with error calculation occurs when a node's interval includes zero, which would require the relative error of the incoming node for (5.42-5.44) to be infinitesimally small. Thus the number of PL segments would go to infinity for such a node. Here the solution might be to give at the beginning some small number as a maximum absolute error of calculation (which makes sense, having in mind the underflow problem with computer calculations). In the process of error calculation when the product of calculated relative error and either of interval boundaries becomes smaller than the given maximum absolute error, the latter is used for the further error calculation.

All the constant's values and variables' intervals in the given example are chosen as to show all the available simplifications. The possible simplifications are always dependent of the overall function's form and all parameter's numerical values. Simplifications are possible during the decomposition as well as during the error assessment. The latter strongly depends on the results of decomposition. The graph on Fig. 5.23 shows only one of a number of possible decompositions. The number of different graphs for an arithmetic expression with n factors and a binary operator is give as

$$\frac{n}{2} \cdot (n-1) \tag{5.45}$$

Table 5.1. Overview of the simplifications for function (5.35) with parameter and variable values (5.36) and (5.37)

Maximum allowable overall relative error	Overall function with simplification
1%	$y = a \cdot x_1^2 \cdot (b \cdot x_2 + c \cdot x_3) \cdot e^{d \cdot x_4}$
4%	$y = a \cdot x_1^2 \cdot (b \cdot x_2 + c_1) \cdot e^{d \cdot x_4}$
32%	$y = a_1 \cdot x_1^2 \cdot e^{d \cdot x_4}$
37%	$y = a_2 \cdot x_1^2$

Full search for all possible configurations, (which is always a finite number), will result in the best simplifications. Table 5.1 depicts the level of simplification for different relative errors, for the given example. The constants a_1, a_2 and c_1 receive their value during the error evaluation procedure.

The simplification done while decomposing the function is the following: when a logarithm (L) and an exponent (E) factor comes one after another both factors can be bypassed. Such pair of nodes is simply omitted in the final implementation. This is the case for factors (y_{11}, y_{14}) and (y_{13}, y_{15}), where y_{10} and y_{12} have the same results as y_{14} and y_{15}, respectively.

When the calculated interval of input values for an L node (the same holds for an E node) appears to be smaller than the product of the node's allowed relative error and either of interval's boundaries, all the branches that lead to that node can be neglected and considered as a constant. For example in Fig 5.23 for e=32%, y_7 is neglected and the branch y_9 becomes constant.

Relative errors propagated from plus and minus nodes allow for yet another simplification of the overall function. The contributions of both incoming branches are evaluated against the allowed relative error of the node. As a result of such evaluation, none, one or both incoming nodes might be neglected or considered as a constant. In the example of Fig. 5.23, factor y_7, for $err = 1\%$ neither of the factors y_5 and y_6 is neglected, for $err = 4\%$ factor y_6 is neglected, for $err = 32\%$ both y_5 and y_6 are neglected, all given errs being overall relative error.

Multiplication of a variable and a constant, as in y_2, y_4, y_5, y_6 and y_{12}, and addition and subtraction, as in y_7, y_{10} and y_{16}, do not require separated PL mappings.

All the necessary data for generating the PL models has been obtained by now: factors or necessary operators, their intervals and allowed relative errors. Each factor gets its PL model that has the minimum number of linear segments for the factor's interval of values and its allowed error, applying Chebyshev approximation. The bold nodes in Fig. 5.23 are implemented (for $err = 1\%$) as separated PL models; other nodes do not require a separate model. The number of nodes that require PL model decreases when the overall relative error increases, which is depicted with the areas of the different shade of gray on Fig. 5.23. This presents a way to trade off the accuracy of the model for the model of smaller size.

The overall PL model is then composed as a hierarchy of factor's PL models, with connectivity information derived from the decomposition process. Such a model is the simplest, and has the minimum number of segments, for the given relative error. The presented method can be embedded within the PL simulator and all the models produced at the run time. This largely reduces the size of the simulator's model library as well.

5.7.2. An example

To show the applicability of such automatic model generator we will again try to model a nMOS transistor described as

$$i_D = 0; \qquad\qquad V_{GS} - V_T \leq 0 \qquad\qquad (5.46)$$

$$i_D = \beta \cdot \left[(V_{GS} - V_T) - \frac{V_{DS}}{2} \right] \cdot V_{DS} \cdot (1 + \lambda \cdot V_{DS}); \; 0 \leq V_{DS} \leq V_{GS} - V_T$$
$$(5.47)$$

$$i_D = \frac{\beta}{2} \cdot (V_{GS} - V_T)^2 \cdot (1 + \lambda \cdot V_{DS}); \qquad 0 \leq V_{GS} - V_T \leq V_{DS}$$
$$(5.48)$$

$$\beta = (K') \cdot \frac{W}{L} \cong (\mu_0 \cdot C_{OX}) \cdot \frac{W}{L} \qquad\qquad (5.49)$$

similar to the ones as defined in (5.15).

These expressions for i_D are indeed scalar multi-variable non-linear functions, as required by the outlined algorithm. The conditions (e.g.. relation between V_T, V_{GS} and V_{DS}) under which these expressions are valid introduce dependency between the input parameters, which can be controlled by means of a "multiplexer switch" that can be implemented as PL model.

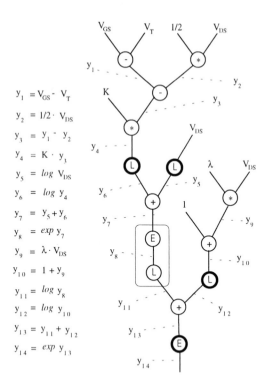

$y_1 = V_{GS} - V_T$

$y_2 = 1/2 \cdot V_{DS}$

$y_3 = y_1 - y_2$

$y_4 = K \cdot y_3$

$y_5 = \log V_{DS}$

$y_6 = \log y_4$

$y_7 = y_5 + y_6$

$y_8 = \exp y_7$

$y_9 = \lambda \cdot V_{DS}$

$y_{10} = 1 + y_9$

$y_{11} = \log y_8$

$y_{12} = \log y_{10}$

$y_{13} = y_{11} + y_{12}$

$y_{14} = \exp y_{13}$

Figure 5.25. Decomposition of linear-region function (5.47)

Figure 5.25 shows straightforward decomposition for the linear region with corresponding factors. This is the result of the parser's decomposition, transformation of multiplications and graph simplification. The PL model requires three logarithm and one exponential PL functions. The factor's intervals are calculated from the values of technology parameters (V_T, λ, K) and expected ranges for input variables (V_G, V_D, V_S). The relative error associated with each factor is calculated from the overall required error, bottom up (see the figure), applying (5.40) to propagate the error from the node's output to its input. The further simplification of the graph is based on the assessment of the relationship between a factor's interval and its assigned relative error. Relative error assigned to each factor depends on the required overall accuracy (the overall relative error) and therefore, depending on the required accuracy some factors might be neglected and omitted from the final model. This provides the efficient way to simplify the formulas (5.47,5.48), for the factor that models channel length modulation can be neglected when the overall required accuracy is smaller than, say, 15%. The model generator will, however, have this phenomenon included into models when the required accuracy is smaller than, say, 15%, and this illustrates model generator's flexibility.

Figure 5.26 shows straightforward decomposition for the saturation region described by (5.48), while Fig. 5.27 shows minimal (from the viewpoint of number of used logarithm and exponential functions) decomposition for the same function.

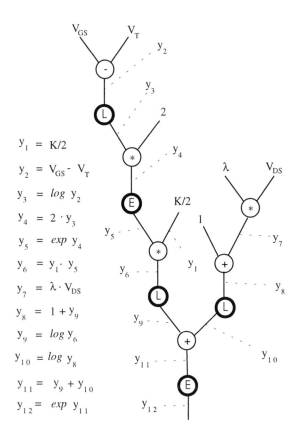

$y_1 = K/2$

$y_2 = V_{GS} - V_T$

$y_3 = log\ y_2$

$y_4 = 2 \cdot y_3$

$y_5 = exp\ y_4$

$y_6 = y_1 \cdot y_5$

$y_7 = \lambda \cdot V_{DS}$

$y_8 = 1 + y_9$

$y_9 = log\ y_6$

$y_{10} = log\ y_8$

$y_{11} = y_9 + y_{10}$

$y_{12} = exp\ y_{11}$

Figure 5.26. Decomposition of saturation-region function (5.48)

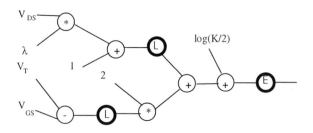

Figure 5.27. Minimal decomposition of function (5.48)

The intervals and errors for the nodes from the Fig. 5.27 are calculated according to the given algorithm. The PL models for the given factors are generated using Chebyshev approximation [1].

To model the transistor's behavior in weak inversion, we can chose among following formulas as the input functions for model generator. The advantage of the proposed technique is that such a PL model has smooth transition between weak inversion and linear region.

The models obtained by the outlined method are assessed by means of simulation. The comparison of the results for the plot of the underlining non-linear function, the generated model, and the previously best manually calculated model as presented in section 5.3 (Fig. 5.28a-c) show superior accuracy of the generated models. The "ideal model" in Fig. (5.28a, 5.29-5.31) refers to plot of the underlining function. The used parameters were: W=20e-6m; L=10e-6m; λ=0.03V^{-1} and K=5.7e-5μA/V^2.

The presented model generator is implemented in C on a Sun Sparc workstation. The generation of the models presented here as examples took well less than 1s CPU time, on a Sparc 10. The required generation time for n-variable function, according to the algorithm, should be $O(n)$.

For 5% relative error the model generator has created a model, shown on Figure 5.28c. This PL model has altogether 43 linear segments for 5% error; factors y_5 and y_6 have 18 segments each, y_{12} has 1 segment and y_{14} has 6 linear segments.

The "old" model form Fig. 5.28 was being used in the PL simulator PLANET [59] and discussed in section 5.3, has 11 linear segments and it was generated by hand, which demanded lots of experience, time and skill. It should be noted here that the old model has reached its limits, namely the modeling of 'multiplication' as for channel length modulation. One can not simply add more segments to get a higher accuracy with the old model.

There are two apparent improvements over the old model: the new model has better accuracy for linear region which, in contrast the old manual model, has the form of the parabolic function (due to more segments) and it better models the effects of channel modulation (due to overall better modeling).

Figure 5.29 shows those flaws of the old model for Vds<1.5V and Vds>3.5V in contrast to Fig. 5.30, which shows the expected accuracy of the new model.

The designer makes the trade-off between model's accuracy and complexity by choosing different overall relative errors, which in turn define the necessary number of linear segments for PL. More accuracy would mean more linear segments that translates into a bigger model and potentially longer simulation times. If we decrease the number of linear segments in order to save computational time, the models' accuracy decreases as well. Figure 5.31 shows the deviation from the underlining functions for the models having 30 and 17 linear segments, respectively and a relative error of 10 and 15%.

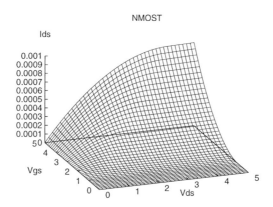

Figure 5.28a. "Ideal" model from underlining functions (5.46-5.49)

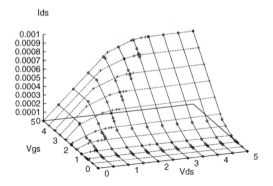

Figure 5.28b: "Old" model from section 5.3.

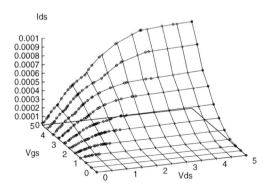

Figure 5.28c. "New" model, generated by PL model generator

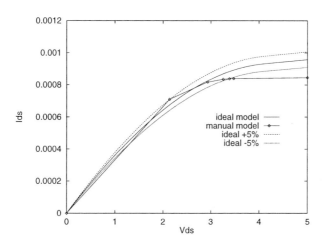

Figure 5.29. Comparison between "old", and "ideal" model

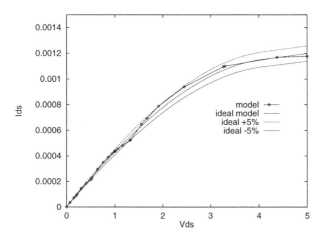

Figure 5.30. Comparison between "new" and "ideal" model

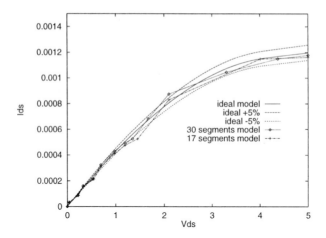

Figure 5.31. "New" model's accuracy-complexity trade-off

5.8.   Concluding remarks

In this chapter, we have demonstrated techniques to generate piecewise linear models for network components on different levels of abstraction, electrical level, logic level and behavioral level. Although it seems sometimes difficult to obtain a PL model for a certain behavior, one has to bear in mind that from a given behavior it is also not obvious how to obtain an analytical expression.

With PL components a network can be built and we discussed simulation techniques to analyze such PL network. PL simulators differ only in their routines to solve time independent equations compared to other network simulators like SPICE. The advantages of PL simulators are in the fact that they are truly MLMS simulators and do have better global convergence properties.

We gave some examples of complex MLMS networks, analyzed by a PL simulator to show the usefulness of such kind of analysis tools.

Further we discussed two examples of automatic PL model generators, starting from analytical expressions. Their advantages are obvious compared to design the models by hand. On the other hand their use is still limited with respect to the complexity of the functions to be handled.

CHAPTER **6**

MULTIPLE SOLUTIONS

So far we discussed methods to obtain the DC operating point of a piecewise linear network. But from circuit theory we know that there also exist circuits having more than one operating point. In this chapter we will discuss several methods to find all operating points of a circuit or more general all solutions of a piecewise linear function.

6.1. Exploiting the lattice structure in Chua's model description

Finding all solutions of a piecewise linear function means solving

$$f(x) = 0 \tag{6.1}$$

In this section we consider f to be according to Chua's canonical model description (see chapter 2),

$$f(x) = a + Bx + \sum_{i=1}^{p} c_i \left| \langle \alpha_i, x \rangle - \beta_i \right| \tag{6.2}$$

Note that (6.2) can be multi-dimensional and that any piecewise linear resistive circuit can be described (6.2).

To obtain all solutions of a piecewise linear function one can use the brute force method. Knowing for each region the linear map $y = \overline{a} + \overline{B}x$, it is easily checked in which region the operating points are and what their values are. However, it means that we have to solve 2^p linear equations that can be a rather large number in the general case. Therefore it is called the brute force way of solving (6.1). Hence it is worthwhile to develop methods which can reduce the computational effort of the task.

In 1982 Chua explored a special property of (6.2) to find all solutions in a more efficient way than the brute force method [82]. This property is the fact that in the domain space all regions are separated by only horizontal and vertical hyperplanes. This is simple due to the absolute value operator as explained in chapter 2. Notice that this property only holds for the one level nested operator. Therefore the method is not applicable for higher order nesting of this operator like in the model description of e.g. Kahlert.

To explain the proposed method, consider the simple circuit shown in Fig. 6.1 where the nonlinear resistors are given by

$$R_{nl1}: i_1 = -\frac{3}{4} + \frac{5}{4}v_1 - \frac{3}{2}|v_1 - 2| + \frac{3}{4}|v_1 - 5| \tag{6.3}$$

$$R_{nl2}: i_2 = \frac{9}{4} + \frac{5}{4}v_2 - \frac{3}{4}|v_2 - 3| \tag{6.4}$$

Further to obey the Kirchhoff laws we have

$$v_1 + v_2 + 2i_1 = 9$$
$$v_1 + v_2 + 2i_2 = 9 \tag{6.5}$$

It is now possible to combine (6.3), (6.4) together with (6.5) into one model, yielding

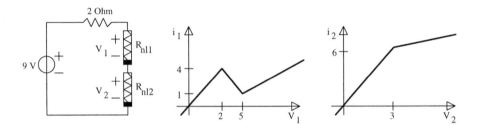

Figure 6.1. A simple piecewise linear circuit

$$f(v_1, v_2) = \frac{1}{4}\begin{pmatrix} -21 \\ -9 \end{pmatrix} + \frac{1}{4}\begin{pmatrix} 7 & 2 \\ 2 & 7 \end{pmatrix}\begin{pmatrix} v_1 \\ v_2 \end{pmatrix} + \begin{pmatrix} -\frac{3}{2} \\ 0 \end{pmatrix}|v_1 - 2| + \begin{pmatrix} \frac{3}{4} \\ 0 \end{pmatrix}|v_1 - 5| + \begin{pmatrix} 0 \\ -\frac{3}{4} \end{pmatrix}|v_2 - 3| = 0$$

$$(6.6)$$

Consider now the domain space of f which is partitioned by 3 hyperplanes into 6 regions, as visualized in Fig. 6.2. Note their special property, they are parallel to one of the axis. Such a structure is called a lattice structure. For each region we have a linear map of f. We can also generate the partitioning of the co-domain or range space by applying map f on the regions in Fig. 6.2 which results in Fig. 6.3.

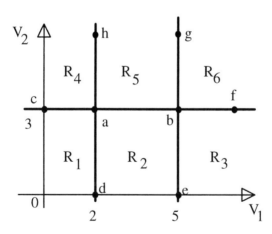

Figure 6.2. The partitions in the domain space of f

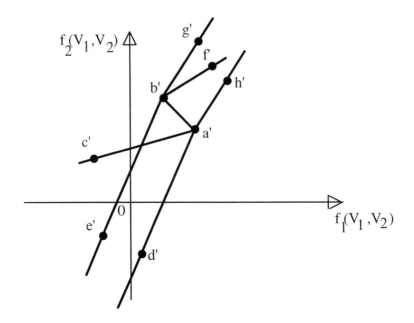

Figure 6.3. The partitions in the range space of f

Because we are searching for the solutions of (6.1) one can see from the range space which regions must be considered, they simply must contain the origin. For our example this results in the regions R_1, R_2 and R_3 because their equivalents in the range space contain the origin. Geometrically this fact can easily be seen, but how can we find those regions in general.

Let x_1 be an arbitrary point in R_2 and let its image be y_1 in $\hat{R}_2$. Consider also hyperplane H_k and its image $\hat{H}_k$

$$H_k: \langle \alpha_k, x \rangle - \beta_k = 0$$
$$\hat{H}_k: \langle \hat{\alpha}_k, x \rangle - \hat{\beta}_k = 0$$

$$(6.7)$$

If the origin is located in region $\hat{R}_2$, then

$$\mathrm{sgn}\left(\langle \hat{\alpha}_k, 0 \rangle - \hat{\beta}_k \right) = \mathrm{sgn}\left(-\hat{\beta}_k \right)$$

$$(6.8)$$

and this must be equal to

$$\text{sgn}\left(\langle \hat{\alpha}_k, y_1 \rangle - \hat{\beta}_k \right) \tag{6.9}$$

because y_1 and the origin must lie on the same side of $\hat{H}_k$. This procedure must be repeated for all sides of the region $\hat{R}_2$. If this so-called *sign test* fails on any of the boundaries of $\hat{R}_2$ then this region contains no solution of (6.1). Due to the sign test we do not have to solve all linear equations, but only those for which we know in advance that they contain a solution. Therefore this method is more elegant than the brute force method. In [82] an efficient implementation of the sign test is described. Let us return to the example. For region R_1 we have as solution

$$f(v) = -\frac{1}{2}\begin{pmatrix} 9 \\ 9 \end{pmatrix} + \frac{1}{2}\begin{pmatrix} 5 & 1 \\ 1 & 5 \end{pmatrix}\begin{pmatrix} v_1 \\ v_2 \end{pmatrix} = 0 \Rightarrow \begin{pmatrix} v_1 \\ v_2 \end{pmatrix} = \frac{1}{2}\begin{pmatrix} 3 \\ 3 \end{pmatrix} \tag{6.10}$$

and in a similar way we obtain the solutions for the regions R_2 and R_3,

$$R_2: \begin{pmatrix} v_1 \\ v_2 \end{pmatrix} = \begin{pmatrix} 4 \\ 1 \end{pmatrix}, \qquad R_3: \begin{pmatrix} v_1 \\ v_2 \end{pmatrix} = \frac{1}{3}\begin{pmatrix} 17 \\ 2 \end{pmatrix} \tag{6.11}$$

We remark that the problem of finding all solutions of a system of piecewise linear (or in general nonlinear) equations is extremely complex. Because a piecewise linear function might have a solution in every region any algorithm which claims to find all solutions must necessarily scan through all possible regions. The efficiency of an algorithm is therefore mainly determined in how efficient it can remove regions that do not have a solution from the list of all regions.

6.2. Separable piecewise linear functions

For finding all solutions of a piecewise linear functions one must find an efficient way to exclude regions that do not contain a solution. The method of Chua performed this by assuming a lattice structure of the PL model and applying a sign test on each boundary of a region. Here we will discuss the algorithm as developed by Yamamura [83].

The important assumption for this method is that one consider the function f to be separable,

$$f(x) = \sum_{i=1}^{n} f^i(x) \tag{6.12}$$

where $f^i : R^1 \rightarrow R^n$. It can be shown that many practical resistive circuits exploit this property and hence this assumption is not too strict [84-85]. Further it is known that a piecewise linear approximation of a separable mapping can be performed on a rectangular subdivision. This means that if f was nonlinear it is transformed into a piecewise linear function by approximating the function linear within each rectangle. Hence a piecewise function will be the result. It means also that the following procedure results in an approximation of the exact solution. The finer the rectangular subdivision, the better the approximated solution of f. If however f was already piecewise linear and especially according to (6.2) than one can choose the subdivision such that it fits with the polytopes of the mapping. In case of (6.2) choose the lattice structure as rectangular subdivision and the exact solutions will be obtained. If this is not possible one can again approximate this piecewise linear function on a chosen rectangular subdivision following the procedure as if the function was nonlinear. So in this sequel we assume f in (6.12) either nonlinear or piecewise linear.

Let us subdivide the solution space into rectangular regions. To this purpose define two vectors

$$l = \left(l_1, l_2, \ldots, l_n\right)^T \quad \text{and} \quad u = \left(u_1, u_2, \ldots, u_n\right)^T \tag{6.13}$$

so that a particular n-dimensional rectangle is given by

$$R = \left\{ x \in R^n \middle| l_i \leq x_i \leq u_i \right\}, \qquad i = 1, 2, \ldots, n \tag{6.14}$$

(see also Fig. 6.4).

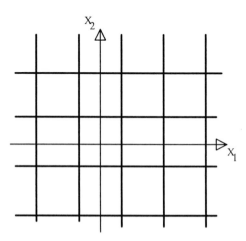

Figure 6.4. Rectangular subdivision of a two dimensional space

Then for this region R define the following sign test

$$\sum_{i=1}^{n}\left[\max\left\{\hat{f}_j^i(l_i), \hat{f}_j^i(u_i)\right\}\right] \geq 0$$

$$\sum_{i=1}^{n}\left[\min\left\{\hat{f}_j^i(l_i), \hat{f}_j^i(u_i)\right\}\right] \leq 0 \qquad j = 1,2,\ldots,n \qquad (6.15)$$

where $\hat{f}$ represents the linear approximation of f in the rectangle under consideration. Equation (6.15) means that in each rectangle only two function evaluations per one region have to be performed. This because the function within the rectangle is linear and hence the function evaluation on the boundaries of the rectangle provides enough information. For instance if we consider the one-dimensional case, then (6.15) reduces to

$$\max\left\{\hat{f}^1(l), \hat{f}^1(u)\right\} \geq 0$$

$$\min\left\{\hat{f}^1(l), \hat{f}^1(u)\right\} \leq 0 \qquad (6.16)$$

which means that at one boundary of the rectangle the function value is positive while at the other boundary the function value is negative. Indeed somewhere within the boundary the function must pass the origin and hence a solution is obtained. If (6.15) does not hold for some j the function does not posses a solution in that rectangle.

This test is very simple, simpler than the one proposed by Chua [82]. Per one region it requires only $2n(n-1)$ additions and $n(n+2)$ comparisons. After the sign test one solves linear equations on the regions that pass the test. The problem with the above outlined method is that the tests have to be applied on each rectangle. We can significantly reduce the number of tests by exploiting another property, namely the *sparsity of the nonlinearity*.

In general each equation is nonlinear or piecewise linear in only a few variables and is linear in the other variables. Suppose that the function f is nonlinear in x_1 and linear in x_2 then we do not have to define a subdivision in R^2 but only in R (see Fig. 6.5). Now we can apply the same sign test of (6.15) to this structure, which has a complexity of a lesser degree than when we had previously. One can show that the total complexity is of order $O(n^3)$.

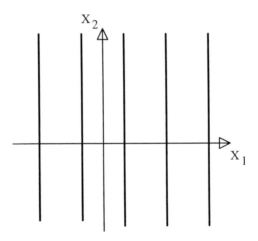

Figure 6.5. New subdivision, exploiting the sparsity of the nonlinearity

Eventually for those regions which pass the test, we can make the subdivision more fine to increase the accuracy in case f is nonlinear.

It can be shown that for those examples for which the function is separable the outlined method is the most powerful. However, we have to keep in mind that if the rectangular subdivision does not fit precisely with the subdivision by the polytopes of a piecewise linear function the solutions of the algorithm are only approximations of the exact solutions. This is a drawback compared to the method of section 6.1 where the exact solutions were obtained.

6.3. Polyhedral methods

In a previous chapter we discussed how a solution algorithm (in this case Katzenelson) can be applied to solve a network of piecewise linear components, i.e. how to find a single operating point for a given excitation. In general terms this means that using a homotopy method one is able to find a single solution of a piecewise linear function starting from an initial condition. Determining all solutions would require trying all possible initial conditions which is a severe drawback. Certainly when the transfer characteristic has unconnected parts, like in the famous example of Fig. 6.6 ([86]).

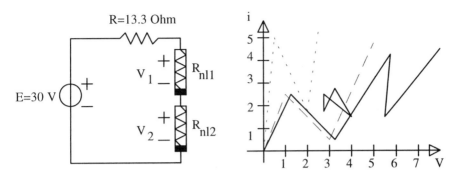

Figure 6.6. Circuit with multiple solutions and two unconnected parts in the transfer characteristic. The dotted curve corresponds to R_{nl1}, the dashed curve to R_{nl2} and the solid curve represents the characteristic for the series network of R_{nl1} and R_{nl2}

We will now discuss two methods that rely on so-called polyhedral techniques. A polytope is described by a set of inequalities. Polyhedral techniques try to find the solution of sets of linear inequalities and give the solution in terms of the corner points of the solution space. This solution space can be obtained by a method proposed by Tschernikow. Therefore we will first outline this method.

6.3.1. Tschernikow's method

The method is developed to find all solutions of the problem

$$Ax \le b, \quad x \in R^n, \quad A \in R^{mxn}, \quad n \ge m \tag{6.17}$$

which in any case with the introduction of some slack variables can be transformed into the problem

$$Bu \le 0, u \ge 0 \quad C \in R^{kxp} \tag{6.18}$$

The solution space of (6.18) describes all non negative solutions of (6.17). The method starts to define a start tableau

$$T^1 = \left(T_1^1 \middle| T_2^1 \right) = \begin{bmatrix} 1 & & 0 & b_{11} & \cdots & b_{k1} \\ & \ddots & & \vdots & & \vdots \\ 0 & & 1 & b_{1p} & \cdots & b_{kp} \end{bmatrix} \tag{6.19}$$

where T_1^1 is a unity matrix, forming a base in the p-dimensional space and T_2^1 is composed by placing a row of (6.18) as column in (6.19). Define for each row in T_1^1 $S(i), i = 1,2,\dots, p$ as the collection of columns in T_1^1 with a zero in row i. Define in a similar way $S(i_1, i_2)$ as the collection with both zeros in i_1 and i_2. Choose now randomly a column in T_2^1, say j, with at least a nonzero element. Consider now two rows, i_1, i_2 from the tableau with opposite sign in column j and consider the corresponding $S(i_1, i_2)$. If now $S(i_1, i_2) \not\subset S(i), i \neq i_1, i \neq i_2$ then the linear combination of rows i_1, i_2 such that a zero in column j is created is of importance. It is just this combination which generates a boundary in the space. Only on one side of this hyperplane, solutions of the problem do exist which are consistent with the space as defined in T_1^1 and the equation as defined by column j which is row j of problem (6.18). Obviously this new row must be introduced in the new tableau matrix. It must be clear that all rows having a zero or negative entry in column j are also transferred to the new tableau matrix. They automatically fulfill the inequality condition in (6.18) for axis j. In a same way tableau T^i can be found from T^{i-1} and the procedure stops when all columns in the right part are treated or one ends up with only columns in the right part which are strict positive. In the latter case there does not exist a solution to the problem except the trivial solution. In the first situation one ends up with the following tableau

$$T^{end} = \left(T_1^{end} \middle| T_2^{end}\right) = \left[\begin{array}{ccc|c} c_{11} & \cdots & c_{1p} & \\ \vdots & & \vdots & 0 \\ c_{t1} & \cdots & c_{tp} & \end{array}\right] \tag{6.20}$$

with the non negative solution

$$u = \sum_{i=1}^{t} p_i c_i, \quad \text{with } c_i = \left(c_{i1}, \dots, c_{ip}\right) \tag{6.21}$$

and p_i a non negative parameter. The set $(c_1, \dots, c_t)^T$ describes the corners of the convex solutions space.

If the problem is written as

$$Ax = b, \quad x \in R^n, \quad A \in R^{mxn}, \quad n \geq m \tag{6.22}$$

then only a small modification in the above outlined procedure is necessary. Only rows having a zero entry in column j are directly transferred to the new tableau matrix. For a detailed outline we refer to the works of Tschernikow [39] or [40,88].

For the generalized LCP the above outlined procedure must only slightly be adapted. Because now the complementarity conditions must also be fulfilled, after each generation of a new tableau we have to check these conditions. We simple check per row over the columns in the left part of the tableau matrix whether they are fulfilled or not. If not, the corresponding row must be removed from the tableau and we can generate a new tableau.

As example consider the following problem

$$2x_1 - 3x_2 + 4x_3 = 0$$
$$- x_1 - 2x_2 + 6x_3 = 0$$

for which the start tableau looks like

$$\begin{pmatrix} 1 & 0 & 0 & 2 & -1 \\ 0 & 1 & 0 & -3 & -2 \\ 0 & 0 & 1 & 4 & 6 \end{pmatrix}$$

Taking the first column of the right hand part, then we can make two combinations of rows having opposite sign. All can be transferred to the next tableau, yielding

$$\begin{pmatrix} 3 & 2 & 0 & 0 & -7 \\ 0 & 4 & 3 & 0 & 10 \end{pmatrix}$$

which finally yields

$$\begin{pmatrix} 30 & 48 & 21 & 0 & 0 \end{pmatrix}$$

resulting in the solution space

$$x = p \begin{pmatrix} 10 \\ 16 \\ 7 \end{pmatrix}$$

and $p \geq 0$. It can easily be verified that this line segment indeed is the cut of the two planes as defined in this example.

We will now outline the procedure to find all solutions of piecewise linear functions.

6.3.2. The method based on Tschernikow's algorithm

In the example of Fig. 6.6 the diodes can be given according to the model description *Bokh2* (see chapter 2), yielding

$$i_1 - 7v_1 + (9.5 \quad -8.5)u = 0$$

$$j = \begin{pmatrix} -1 \\ -1 \end{pmatrix} v_1 + Iu + \begin{pmatrix} 0.7 \\ 2.05 \end{pmatrix} \qquad\qquad (6.23)$$

$$i_2 - 2.5v_2 + (3.5 \quad -3)u = 0$$

$$j = \begin{pmatrix} -1 \\ -1 \end{pmatrix} v_2 + Iu + \begin{pmatrix} 1 \\ 3.15 \end{pmatrix} \qquad\qquad (6.24)$$

where the complementary conditions are left out for convenience.

In [38,88] a method is proposed to obtain all solutions of a piecewise linear function like (6.23) and (6.24) where eventually there may be restrictions on the input and output parameters. The method consists of 3 steps if we leave out the restrictions on the input and output variables. We will go through the steps at the hand of our example.

In step 1 we try to find an exact description of each polytope. We know that each polytope can be given by $C_{i \bullet} x + g_i \leq 0$ with i from the set of all hyperplanes. This means that we have to solve a set of inequalities of the form

$$Fu \leq 0, u \geq 0 \qquad\qquad (6.25)$$

where a slack variable is used to handle g_i. Note that (6.25) is under constrained and hence generate a solution space instead of a single solution point. Such a set can be solved by the methods as proposed by Tschernikow and described in the previous sub section. The obtained vectors of (6.25) exactly define the corner points of the polytope and hence each positive linear combination of these vectors define a point within the polytope. Label the polytopes for which a solution does exist and proceed with step 2.

Find for each labeled polytope the corresponding linear mapping, for instance by using Katzenelson or simple by pivoting over the corresponding hyperplanes for that polytope. Combine the linear mapping with the solution as found in step 1, transform the set of equations into a set according (6.27) and solve this set. For the piecewise linear resistors (6.25) yields

$$\begin{pmatrix} i_1 \\ v_1 \end{pmatrix} = p_{11} \begin{pmatrix} 0 \\ 0 \end{pmatrix} + p_{12} \begin{pmatrix} 4.9 \\ 0.7 \end{pmatrix}$$

$$\begin{pmatrix} i_1 \\ v_1 \end{pmatrix} = p_{21} \begin{pmatrix} 1.53 \\ 2.05 \end{pmatrix} + p_{22} \begin{pmatrix} 4.9 \\ 0.7 \end{pmatrix} \qquad (6.26)$$

$$\begin{pmatrix} i_1 \\ v_1 \end{pmatrix} = p_{31} \begin{pmatrix} 1.53 \\ 2.05 \end{pmatrix} + \begin{pmatrix} 6 \\ 1 \end{pmatrix}$$

$$p_{11} + p_{12} = 1, p_{21} + p_{22} = 1, \forall_{ij} p_{ij} \geq 0$$

and a similar set is obtained for (6.24). Equation (6.26) simple describes all segments of the piecewise linear characteristic as given in Fig. 6.6. One can now simple combine this with the Kirchhoff laws and find the solutions of Table 6.1.

In the case the components of the circuit are one-dimensional, this procedure is far more complex than the one using the generalized LCP (see next section). However, when more dimensional characteristics are used that method is not usable and one can rely on this one. There are some other advantages of this procedure.

First the method can be used in a hierarchical environment. This means that solutions obtained on a lower level can be used to solve equations on a higher level. In our example we first solved the problem on the level of the resistors, later on we try to fit these solutions with the solutions obtained from obeying the Kirchhoff laws. One can extend this towards large systems built up from a set of sub systems. First all solutions for each sub system are obtained. Later, when the complete system is built, one easily obtains the overall solutions space by combining the obtained solutions of the sub systems according to the Kirchhoff laws.

Secondly, any PL function of any dimension can be handled. This is in contrast to the methods of Chua and Yamamura. In the first case the individual elements must be one-dimensional and must be described according to the model description *Chua*. In chapter 2 we saw that many characteristics can not be modeled in this format. Besides we always end up with an *n*-by-*n* system, so no degeneracy is allowed. Using the technique of Tschernikow, under constrained problems can be handled as well. In the case of Yamamura the assumption of separable functions makes it difficult to analyze multi-dimensional functions for which we do not know if they fulfill this condition.

Finally, the advantage of methods based on the Tschernikow algorithm is that they do not need any starting point and are exact, i.e. they find the exact values of all solutions.

Table 6.1. Nine solutions for the circuit in Fig. 6.6.

V_1	0.31	0.30	0.28	1.84	1.86	1.93	2.14	2.13	2.10
V_2	0.87	1.37	3.94	0.82	1.50	3.88	0.80	1.52	3.88

6.3.3. The method based on the Generalized LCP

Any one dimensional PL mapping can be written according to

$$x = x_0 + x_{-\infty}\lambda^- + (x_1 - x_0)\lambda^+ +$$

$$\sum_{k=2}^{n}(x_k - 2x_{k-1} + x_{k-2})\lambda_{k-1}^+ + (x_{+\infty} - 2x_n + x_{n-1})\lambda_n^+ \qquad (6.27)$$

$$\lambda_j^+ - \lambda_j^- = \lambda^+ - \lambda^- - j, \quad j = 1,2,\ldots,n$$

$$\lambda_j^+, \lambda_j^-, \lambda^+, \lambda^- \geq 0 \qquad (6.28)$$

$$\lambda^+ \cdot \lambda^- = 0, \lambda_j^+ \cdot \lambda_j^- = 0$$

where $x_j, j = 1,2,\ldots,n$ represents a breakpoint in the characteristic and $x_{-\infty}, x_{+\infty}$ represents some points at the most left and most right segment [87]. Relation (6.27) describes a PL mapping with the parameters consistent with the complementarity conditions as given in (6.28). A geometrical interpretation of (6.27) is given in Fig. 6.7.

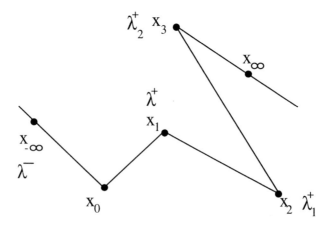

Figure 6.7. Geometrical interpretation of (6.27).

For our example the two nonlinear resistors (Fig 6.6) can then be represented by

$$\begin{pmatrix} i_1 \\ v_1 \end{pmatrix} = \begin{pmatrix} 4.9 \\ 0.7 \end{pmatrix} + \begin{pmatrix} -4.9 \\ -0.7 \end{pmatrix} \lambda^- + \begin{pmatrix} -3.37 \\ 1.35 \end{pmatrix} \lambda^+ + \begin{pmatrix} 6.07 \\ -0.9 \end{pmatrix} \lambda_1^+$$

$$\lambda_1^+ - \lambda_1^- = \lambda^+ - \lambda^- - 1 \tag{6.29}$$

$$\lambda^+, \lambda^-, \lambda_1^+, \lambda_1^- \geq 0$$

$$\lambda_1^+ \cdot \lambda_1^- + \lambda^+ \cdot \lambda^- = 0$$

and

$$\begin{pmatrix} i_2 \\ v_2 \end{pmatrix} = \begin{pmatrix} 2.5 \\ 1 \end{pmatrix} + \begin{pmatrix} -2.5 \\ -1 \end{pmatrix} \mu^- + \begin{pmatrix} -1.9 \\ 2.15 \end{pmatrix} \mu^+ + \begin{pmatrix} 5.6 \\ -0.3 \end{pmatrix} \mu_1^+$$

$$\mu_1^+ - \mu_1^- = \mu^+ - \mu^- - 1 \tag{6.30}$$

$$\mu^+, \mu^-, \mu_1^+, \mu_1^- \geq 0$$

$$\mu_1^+ \cdot \mu_1^- + \mu^+ \cdot \mu^- = 0$$

The linear circuit equations are

$$E - Ri_1 - (v_1 - v_2) = 0$$
$$i_1 - i_2 = 0 \tag{6.31}$$

and can be rewritten as

$$\begin{pmatrix} 1 & 0 & -1 & 0 & 0 \\ -R & -1 & 0 & 1 & E \end{pmatrix} \begin{pmatrix} i_1 \\ v_1 \\ i_2 \\ v_2 \\ \alpha \end{pmatrix} \tag{6.32}$$

where α slack parameter. We can now substituting (6.29) and (6.30) into (6.32) yielding a system of the following form

$$Mw + Nz = q\alpha$$
$$w, z \geq 0, \alpha \geq 0 \tag{6.33}$$
$$w^T \cdot z = 0$$

which in the literature is known as the *Generalized Linear Complementarity Problem* [37,87]. The set can be transformed into

$$\begin{pmatrix} M & N & -q \end{pmatrix} \begin{pmatrix} w \\ z \\ \alpha \end{pmatrix} = 0 \tag{6.34}$$

with $w, z, \alpha \geq 0$ and the complementarity condition still valid. This set of equations can be solved using the modified Tschernikow method as shortly discussed in section 3.5 and outlined in the preceding section. The modification is due to the fact that one has to take into account the complementarity condition. Note that the term 'generalized' is used because the matrix is not of dimension $R^{n \times n}$ as like the LCP as discussed in chapter 3 but can have any dimension, i.e. $R^{n \times m}, n \leq m$. Hence it can represent an under constrained set of equations which indeed can posses more than one solution.

Solving (6.34) for the example will yield nine solutions as they are given in Table 6.1.

So the main flow of the outlined method is to describe all one-dimensional PL functions according to the format (6.27, 6.28). The descriptions can then be used in the Kirchhoff laws, resulting in a system like (6.33). This system can then be solved, resulting in obtaining all solutions of the problem.

Finally we will give an other network example. The nonlinear resistor as defined by the network in Fig.6.8.can be given in terms of (6.27-6.28) as

$$\begin{pmatrix} i \\ v \end{pmatrix} = \begin{pmatrix} 1 \\ 1 \end{pmatrix} + \begin{pmatrix} -1 \\ -1 \end{pmatrix} \lambda^- + \begin{pmatrix} -\frac{1}{2} \\ 1 \end{pmatrix} \lambda^+ + \begin{pmatrix} \frac{3}{2} \\ 0 \end{pmatrix} \lambda_1^+$$
$$\lambda_1^+ - \lambda_1^- = \lambda^+ - \lambda^- - 1 \tag{6.35}$$
$$\lambda^+, \lambda^-, \lambda_1^+, \lambda_1^- \geq 0$$
$$\lambda_1^+ \cdot \lambda_1^- = \lambda^+ \cdot \lambda^- = 0$$

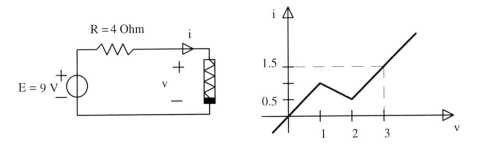

Figure 6.8. The network and the piecewise linear characteristic of the resistor.

and the topological equation (6.31) can be rewritten as

$$
\begin{pmatrix} -6 & -1 & 6 \end{pmatrix} \begin{pmatrix} i \\ v \\ \alpha \end{pmatrix}
\tag{6.36}
$$

where α is a slack parameter. We can now substitute (6.35) into (6.36) yielding a system of the following form

$$
\begin{pmatrix} -9 & 0 & 2 & 7 & -1 \\ -1 & 1 & 1 & -1 & -1 \end{pmatrix} \begin{pmatrix} \lambda_1^+ \\ \lambda_1^- \\ \lambda^+ \\ \lambda^- \\ \alpha \end{pmatrix}
\tag{6.37}
$$

The start tableau looks like

$$
\begin{pmatrix}
1 & 0 & 0 & 0 & 0 & -9 & -1 \\
0 & 1 & 0 & 0 & 0 & 0 & 1 \\
0 & 0 & 1 & 0 & 0 & 2 & 1 \\
0 & 0 & 0 & 1 & 0 & 7 & -1 \\
0 & 0 & 0 & 0 & 1 & -1 & -1
\end{pmatrix}
$$

Taking the first column of the right hand part, we can make two combinations of rows having opposite sign. All can be transferred to the next tableau, yielding

$$\begin{pmatrix} 0 & 1 & 0 & 0 & 0 & | & 1 \\ 2 & 0 & 9 & 0 & 0 & | & 7 \\ 7 & 0 & 0 & 9 & 0 & | & -16 \\ 0 & 0 & 1 & 0 & 2 & | & -1 \\ 0 & 0 & 0 & 1 & 7 & | & -8 \end{pmatrix}$$

which finally yields (because many combinations do not fulfill the complementary conditions)

$$\begin{pmatrix} 0 & 1 & 1 & 0 & 2 \\ 2 & 0 & 16 & 0 & 14 \\ 0 & 8 & 0 & 1 & 7 \end{pmatrix} \tag{6.38}$$

We now consider the first equation in (6.38) which tells us that $\lambda_1^+ = \frac{1}{2}$ ($\alpha \equiv 1$). Combining this with (6.35) leads to $(i,v) = \left(\frac{3}{4}, \frac{3}{2}\right)$ which is indeed one of the DC-operating points. In a similar approach the other two operating points can be obtained from (6.38), $(i,v) = \left(\frac{6}{7}, \frac{6}{7}\right)$ and $(i,v) = \left(\frac{9}{14}, \frac{15}{7}\right)$.

6.4. Polyhedral methods and Linear Programming

For a long time the relation between the LCP and Linear Programming (LP) is known. Each LP problem can be transformed into an LCP using the duality property of the LP. We mentioned this property already in chapter 3. On the other hand it is possible to treat a piecewise linear network as a polyhedral function which can then be solved using LP. We mentioned that the state equation describes a set of polyhedral regions in the space, called polytopes. For each polytope a linear relation describes the local behavior of the function. We can also combine these two relations when we treat the piecewise linear function as a polyhedral element. The polyhedral elements, in general, do not have a correspondence with a physical device, but they constitute a mathematical tool. Each polyhedral element consists of a set of polyhedral regions. For our nonlinear resistor in Fig. 6.8 one of the polyhedral sections would yield

$$\begin{pmatrix} i \\ v \end{pmatrix} = p_1 \begin{pmatrix} 0 \\ 0 \end{pmatrix} + p_2 \begin{pmatrix} 1 \\ 1 \end{pmatrix} + p_3 \begin{pmatrix} \frac{1}{2} \\ 2 \end{pmatrix} = \sum_{i=1}^{3} p_i w_i$$

$$p_1 + p_2 + p_3 = 1 \qquad\qquad\qquad\qquad\qquad\qquad\qquad\qquad (6.39)$$

$$p_1, p_2, p_3 \geq 0$$

describing the triangular area defined by the first two segments and a virtual line segment. We can use this description together with the topological equations to obtain a set similar to

$$\sum_{m=1}^{M} \sum_{i=1}^{K_m} p_i^m \left(w_i^m t_{km} \right) = r_k \quad \text{for } k = 1,\ldots,M \qquad\qquad (6.40)$$

$$\sum_{i=1}^{K_m} p_i^m = 1, p_i^m \geq 0$$

where t and r define the topological relations, M is the number of polyhedral elements in the network and K_m the number of polyhedral regions per element. This system of linear equations and inequalities may be regarded as the constraints of a Linear Programming problem. The solution for each polyhedral region is a DC-operating point for the original problem. To find all operating points implicates that in principle all polyhedral regions have to be solved. However, when one sets up a 'genealogical tree', a reasonable reduction in computation can be obtained. A certain node in this tree represents a specific polyhedral element. If a certain node does not contain a solution some other nodes in the tree may be discarded. In [93] a detailed discussion can be found.

6.5.   Tracing the driving point and transfer characteristic

Often a simulator is used to trace the driving point characteristic of a system. When a system does posses only one closed solution curve there will be no problem, when we rely on the methods as discussed in chapter 4. If however the characteristic contains several disjunct curves, often resulting in several operating points when the system is loaded, we might have a problem. We can now consider two situations:

1. Suppose that we do know for each disjunct part of the characteristic an initial condition. Then to find the characteristic we simple can rely on the methods as proposed by e.g. Katzenelson,... We restart the solution procedure for each initial condition and we end up with the complete transfer characteristic or driving point curve. If additional the PL models possesses the lattice structure we can use the method proposed in [89]. Due to the lattice structure one is able

to find the breakpoints in the characteristics easily and explicitly. Therefore the method is extremely fast. Problems as the discussed example of Fig. 6.6 can be solved easily.

2. Under point 1 we solved the problem in two steps. First find all solutions and then use them as starting points to trace the curves. However this method fails if we do not know initial conditions. Hence by trial and error we have to find initial conditions on each disjunct curve, where we might even not know how many disjuct curves there are. So far known the only two methods to overcome this problem are the use of the generalized LCP and the polyhedral method. These two algorithms do not need any initial starting point and find all solutions and implicitly the complete driving point or transfer characteristic. See for instance solution (6.18) which is the complete driving point characteristic of the resistor. At a higher level, the complete characteristic of the two resistors together is obtained (see Fig. 6.6) and in the last step by using the load of the circuit, the operating points.

6.6. Some remarks

We recall that the problem of finding all solutions of a system of piecewise linear (or in general nonlinear) equations is extremely complex. A piecewise linear function might have a solution in every region. Therefore any algorithm which claims to find all solutions must necessarily scan through all possible regions. The efficiency of an algorithm is therefore mainly determined in how efficient it can remove regions that do not have a solution from the list of all regions.

We discussed several algorithms each having advantages and drawbacks. Except for the algorithms that are based on techniques to find all solutions of linear (in) equalities, all methods assume property of the piecewise linear function. This can be a lattice structure of the space or a separable function property. Obviously if they can be applied, such algorithms are faster than the two general methods.

BIBLIOGRAPHY

[1] E.W. Cheney, *Introduction to approximation theory*, McGraw Hill, 1966

[2] W.M.G. van Bokhoven, *Piecewise Linear Modelling and Analysis,* Deventer, The Netherlands: Kluwer Technische Boeken, 1981

[3] L.O. Chua, S.M. Kang, "Section-wise piecewise linear functions: Canonical representation, properties and applications," *Proc. IEEE*, vol. 65, pp. 915-929, June 1977

[4] W.M.G. van Bokhoven, "Piecewise Linear analysis and simulation," in *Circuit Analysis, Simulation and Design*, A.E. Ruehli (Ed.). Amsterdam: North-Holland, 1986, Ch.9.

[5] C.E. Lemke, "On the complementary pivot-theory," in *Mathematics of decision sciences*, part I, Eds. G.B. Dantzig and A.F. Veinott Jr. New York, Academic Press, 1970

[6] S.M. Kang, L.O. Chua, "A global representation of multi-dimensional piecewise linear functions with linear partitions," *IEEE Trans. Circuits Syst.*, vol. 25, Nov., pp.938-940, 1978

[7] L.O. Chua, A.C. Deng, "Canonical piecewise linear representation," *IEEE Trans. Circuits Syst.*, vol. 35, Jan., pp.101-111, 1988

[8] G. Güzelis, I. Göknar, "A canonical representation for piecewise affine maps and its application to circuit analysis," *IEEE Trans. Circuits Syst.*, vol. 38, Nov., pp.1342-1354, 1991

[9] C, Kahlert, L.O. Chua, "A generalized canonical piecewise linear representation," *IEEE Trans. Circuits Syst.*, vol. 37, March, pp.373-382, 1990

[10] C, Kahlert, L.O. Chua, "The complete canonical piecewise-linear representation-part I: The geometry of the domain space," *IEEE Trans. Circuits Syst., part-I,* vol. 39, March, pp.222-236, 1992

[11] J.L. Huertas, A. Rueda, A. Rodriguez-Vaguez, "Multi-dimensional piecewise linear partitions: global canonical representation," *Proc. ISCAS*, pp. 1106-1109, 1984

[12] T.A.M. Kevenaar, D.M.W. Leenaerts, "A comparison of piecewise-linear model descriptions," *IEEE Trans. Circuits Syst., part-I,* vol. 39, Dec., pp.996-1004, 1992

[13] T.A.M. Kevenaar, D.M.W. Leenaerts, W.M.G. van Bokhoven, "Extensions to Chua's explicit piecewise linear function descriptions," *IEEE Trans. Circuits Syst., part-I,* vol. 41, April, pp.308-314, 1994

[14] D.M.W. Leenaerts, "Further Extensions to Chua's Explicit Piecewise Linear Function Descriptions," *Int. Journal of Circuit Theory and Appl.*, vol. 24, pp.621-633,1996.

[15] J.T.J. van Eijndhoven, "Piecewise linear analysis" in *Analog methods for computer-aided circuit analysis and diagnosis,* T. Ozawa (ed.), Marcel Dekker Inc. 1988, New York, Chap.3

[16] H.W. Buurman, *From circuit to signal, development of a piecewise linear simulator,* Ph.D. dissertation, Technical University of Eindhoven, 1993

[17] A.N. Kolmogorov, "On the representation of continuous functions of many variables by superposition of continuous functions of one variable and addition," *Dokl. Akad. Nauk. USSR, 114, 1957, pp.953-956*

[18] D.A. Sprecher, "On the structure of continuous functions of several variables," *Trans. of the American Mathematical Society,* vol. 115, pp. 340-355, 1965

[19] K. Funahashi, "On the approximate realization of continuous mappings by neural networks," *Neural Networks,* vol.2 pp. 183-192, 1989

[20] D.M.W. Leenaerts, "On the relation between neural networks and the LCP," proc. ECCTD'95, Istanbul, Turkey, pp.577-580 1995

[21] C.E. Lemke, "On complementary pivot theory," *Mathematics of the Decision Sciences. Part 1* Proc. 5yth Summer Seminar, Standford Calif. 10 July-11 Aug. 1967, Ed. by G.B. Dantzig and A.F. Veinott, Jr. Providence, R.I.: American Mathematical Society, 1968. Lectures in Applied Mathematics, vol. 11. pp. 95-114.

[22] J. Katzenelson, "An algorithm for solving nonlinear resistor networks," *Bell Syst.Tech. J.* vol. 44, 1965, pp.1605-1620

[23] C. van de Panne, "A complementary variant and a solution algorithm for piecewise linear resistor networks," *SIAM J. Math. Anal.* vol. 8 1977, pp.69-99

[24] D.M. Wolf, S.R. Sanders, "Multiparameter homotopy methods for finding DC operating points of nonlinear circuits," *IEEE Trans. Circuits and Syst. part I,* vol. 43, no. 10, pp 824-838, 1996

[25] C.W. Cryer, "The solution of a quadratic programming problem using systematic overrelaxation," *SIAM J. of Contr.* vol 9, no. 3 1971, pp.385

[26] M. Fiedler, V. Pták, "Some generalizations of positive definiteness and monotonicity" *Num. Math.* vol. 9, no 12, 1966 pp.163-172

[27] S. Kamardian, "The complementarity problem," *Math. Program.,* vol.2 1972, pp.107-129

[28] C.E. Lemke, "Bimatrix equilibrium points and mathematical programming," *Manage. Sci.* vol. 11 1965, pp.681-689

[29] L.B. Rall, *Computational solution of nonlinear operator equations,* J. Wiley & Sons Inc. New York, 1969, pp. 65.

[30] R.W. Cottle, G.B. Dantzig, "Complementary pivot theory of mathematical programming," *Linear Algebra & Appl.* vol. 1, 1968 p.103-125

[31] B.C. Eaves, "The linear complementary problem," *Manage. Sci.* vol. 17, 1971 p. 225-237

[32] J.T.J van Eijndhoven, *A piecewise linear simulator for large scale integrated circuits,* Ph.D. thesis Technische Hogeschool Eindhoven, the Netherlands, 1984

[33] M.J. Chien, E.S. Kuh, "Solving piecewise linear equations for resistive networks," *Int. Journal of Circuit Theory and Appl.*, vol. 4, pp. 3-24., 1976

[34] G. van der Laan, A.J.J. Talman," Convergence and properties of recent variable dimension algorithms," *Numerical Solution of Highly Nonlinear Problems: Fixed point algorithms and complementary problems.* Ed. W. Forster, Amsterdam, North-Holland, 1980, pp.3-36

[35] G. van der Laan, "Simplicial fixed point algorithms," Amsterdam: Mathematisch centrum, 1980, Mathematical Centre Tracts, 129

[36] A.J.J. Talman, "Variable dimension fixed point algorithms and triangulations," Amsterdam: Mathematisch cetrum, 1980, Mathematical Centre Tracts, 128

[37] B. de Moor, *Mathematical concepts and techniques for modelling of static and dynamic systems*, Ph.D. Thesis Katholieke Universiteit Leuven, Belgium, 1988.

[38] D.M.W. Leenaerts, J.A. Hegt, "Finding all solutions of piecewise linear functions and the application to circuit design," *Int. Journal Circuit Theory and Appl.*, vol. 19, 1991, pp.107-123

[39] S.N. Tschernikow, *Lineare Ungleichungen* Berlin:VEB Deutscher Verlag der Wissenschaft, 1971 (translation form: *Lineinye neravenstva,* 1968 by H. Weinert and H. Hollatz into German)

[40] D.M.W. Leenaerts "Applications of interval analysis to circuit design," *IEEE trans. Circuits and Syst.,* vol. CAS-37, no. 6, 1990, pp.803-807

[41] S. Polak, W. Schilders, M. Dressen, "The CURRY algorithm" in *Simulation of semiconductor devices and processes*, vol 2. Pineridge Pres Swansea, 1986, pp.131-177

[42] S. Selberherr, *Analysis and simulation of semiconductor devices* Springer-Verlag, New York, 1984

[43] L.W. Nagel, "SPICE2: A computer program to simulate semiconductor circuits," *Electr. Res. Lab. Rep.* No. ERL-M520, Univ. of California, Berkeley, May 1975

[44] Vladimirescu, S. Liu, "The simulation of MOS integrated circuits using SPICE 2,"*Memorandum* No. UCB/ERL M80/7, Electronics Research Laboratory, College of Eng. University of Berkeley, Febr. 1980

[45] K.A. Sakallah, S.W. Director, "Samson2, An event driven VLSI circuit simulator," *IEEE Trans. Computer-Aided Design* CAD-4 (4) 1985, pp. 668-684

[46] "Advanced statistical analysis program (ASTAP)," IBM Corp. Data Proc. Div., White Plains, NY, Pub. No. SH20-1118-0

[47] W.T. Weeks et.al. "Algorithms for ASTAP- a network analysis program," *IEEE Trans. Circuit Theory* vol. CT-20, No. 6. pp. 628-634

[48] R.E. Bryant, et.al. "COSMOS:a compiled simulator for MOS circuits," proc 24th ACM/IEEE Design Automation Conf. 1987, pp.9-16

[49] T.H. Krodel, K.J. Antreich, "An accurate model for ambiguity delay simulation," proc. European Design Automation, 1990, pp. 563-567

[50] R.E. Bryant, "A switch-level model and simulator for MOS digital systems," *IEEE Trans.Computers*, vol. C-33, no.2 pp.160-177, 1984

[51] A.J. van Genderen, A.C. de Graaf, "SLS: a switch level timing simulator," in *The Integrated Design Book - papers on VLSI Design Methodology from the ICD-NELSIS project*, ed. P. Dewilde, pp.2.93-2.146 Delft University Press, 1986

[52] S.M. Shah, et. al. "MATRIXx control design and model building" in *Computer-Aided Control systems engineering*, ed. M. Janshidi, C.J. Herget, Elsevier Science Publishers B.V. Amsterdam, 1985, pp.181-207

[53] IEEE Standard VHDL language Reference Manual, IEEE std. 1076-1987 IEEE New York 1988

[54] Saber, user manual, Analogy, Inc.

[55] FIDELDO, Anacad, Inc.

[56] SMASH, user manual

[57] M.T. van Stiphout, et. al. "PLATO: a new piecewise linear simulation tool," proc. European Design Automation Conf. pp. 235-239, 1990

[58] L.O. Chua, "Canonical piecewise linear analysis: part II-tracing point and transfer characteristics," *IEEE Trans on Circuits and Syst.* Vol. CAS-32, May 1985, pp. 417-433

[59] T.A.M. Kevenaar, D.M.W. Leenaerts, "A flexible hierarchical piecewise linear simulator," *Integration, the VLSI journal*, Vol. 12 1991, pp. 211-235

[60] J. Vlach, K. Singhal, *Computer methods for analysis and design*, Van Nostrand Reinhold, New York, 1983, chapter 13

[61] C.W. Gear, *Numerical initial value problems in ordinary differential equations* Prentice hall, Englewood Cliffs, NJ. 1971

[62] G. Dahlquist, "A special stability problem for linear multistep methods," *BIT*, Vol. 3, pp.27-43, 1963

[63] R.K. Brayton, "Error estimates for the variable-step backward differentiation methods," IBM Research Report RC 6205 (#26655) September 1976

[64] R.K. Brayton, F.G. Gustavson, G.D. Hachtel, "A new efficient algorithm for solving differential-algebraic systems using implicit backward differentiation formulas," *Proc. IEEE*, Vol. 60, No. 1, pp.98-108, January 1972

[65] W.M.G. van Bokhoven, "Linear implicit differentiation formulas of variable step and order," *IEEE Trans. Circuits Syst.* Vol. CAS-22, No. 2, pp. 109-115, February, 1975

[66] W.M.G. van Bokhoven, "An activity controlled modified waveform relaxation method," *Proc. ISCAS*, New York, 1983, pp. 766-768.

[67] T.A.M. Kevenaar, "Periodic steady state analysis using shooting and wave-form-Newton," *Int. Journal Circuit Theory and Appl.*, Vol. 22, 1994, pp.51-60

[68] C. Runge, "Uber die numerische Auflösung von Differential Gleichungen," *Math. Ann* vol. 46, pp. 167-178, 1895

[69] W. Kutta, "Beitrag zur naherungs weisen Integration von Differential Gleichungen," *Zeit. Math Physik*, vol. 46, pp.435-453, 1901

[70] J. Stoer, R. Bulirsh, *Introduction to numerical analysis,* Springer-Verlag, New York, 1980

[71] S.W. Director, R.A. Rohrer, "The generalized adjoint network and network sensitivities," *Trans Circuits and Syst.*, Vol. CT-26, no.3 August 1969

[72] D.A. Clahan, *Computer-aided network design*, McGraw-Hill Inc. 1972

[73] A.F. Schwarz, *Computer aided design of microelectronic circuits and systems,* Academic Press, 1987, London

[74] L.O. Chua, A. Deng, "Canonical piecewise linear modeling," *IEEE Trans. Circuits and Syst.* Vol. CAS- 33, No. 5 May, 1986, pp.511-525

[75] M.L. Dertouzos, *Threshold logic: a synthesis approach*, MIT press, 1965

[76] S.L. Hurst, *Threshold logic*, Mill & Boon Ltd., London 1971

[77] T.A.M. Kevenaar, *PLANET, a hierarchical network simulator* Ph.D. dissertation, Eindhoven, The Netherlands, 1993

[78] High speed CMOS, Philips data handbook, IC06N, pp.398-402, 1986

[79] Z.Y. Chang, W.M.C. Sansen, *Low-noise, wide-band amplifiers in Bipolar and CMOS technologies*, Boston Kluwer, Academic Publishers, 1991

[80] W. Kruiskamp, D. Leenaerts, "A CMOS peak detect sample and hold circuit, " *IEEE Trans. Nucl. Science,* vol. 41, no.1 pp.295-298, Feb 1994

[81] W. Kruiskamp, D. Leenaerts, "Behavioral and macro modeling using piecewise linear techniques," *Int. Journal on Analog Integrated Circuits and Signal Processing*, Vol. 10, no. 1/2, pp.67-76, 1996

[82] L.O. Chua, R.L.P. Ying, "Finding all solutions of piecewise-linear circuits," *Int. Journal Circuit theory and Appl.* Vol. 10, pp.201-229, 1982

[83] K. Yamamura, "Finding all solutions of piecewise-linear resistive circuits using simple sign tests," *IEEE Trans. Circuits and Syst.* Vol. 40, pp. 546-551, 1993

[84] K. Yamamura, M. Ochiai, "An efficient algorithm for finding all solutions of piecewise-linear resistive circuits," *IEEE Trans. Circuits and Syst.* Vol. 39, pp. 213-221, 1992

[85] K. Yamamura, "Exploiting separability in numerical analysis of nonlinear systems," *IEICE Trans. Fundamentals of Electron., Commun., Comput. Sci.,* vol. E75-A, pp.285-293, 1992

[86] L.O. Chua , "A switching parameter algorithm for finding multiple solutions of nonlinear resistive functions," *Int. Journal Circuit theory and Appl.* Vol. 4, pp.215-239, 1976

[87] L. Vandeberghe, B. De Moor, J. Vandewalle, "The generalized linear complementarity problem applied to the complete analysis of resistive piecewise linear circuits," *IEEE Trans. Circuits and Syst.* Vol. 36, pp. 1382-1391, 1989

[88] D.M.W. Leenaerts, *TOPICS a contribution to design automation* Ph.D. dissertation Dep. of Elect Eng., Technical University Eindhoven, 1992

[89] L.O. Chua, A. Deng, "Canonical piecewise linear analysis: part II-tracing driving-point and transfer characteristics," *IEEE Trans. Circuits and Syst.* Vol. 32, pp. 417-444, 1985

[90] P. Veselinovic, D.M.W. Leenaerts, "A method for automatic generation of piecewise linear models," Proc. ISCAS, Atlanta, part III, pp.24-27, 1996

[91] P. Pejovic, D. Maksimovic, "An algorithm for solving piecewise linear networks that include elements with discontinuous characteristics," *IEEE Trans. Circuits and Syst. Part I.* Vol. 43, pp. 453-460, 1996

[92] K. Kawakita, T. Ohtsuki, "NECTAR 2 a circuit analysis program based on piecewise linear approach," *Proc. ISCAS* Boston, pp.92-95, 1975

[93] S. Pastore and A. Premoli, 1993, Polyhedral elements: a new algorithm for capturing all equilibrium points of piecewise linear circuits, *IEEE Trans. Circuits Syst. I, Fundam. Theory Appl. Vol.* 40: 124-132, 1993

INDEX